The Evolution of Faith

ALSO BY PHILIP GULLEY

If the Church Were Christian
I Love You, Miss Huddleston
If Grace Is True
If God Is Love
Porch Talk
Front Porch Tales
Hometown Tales
For Everything a Season
Home to Harmony
Just Shy of Harmony
Signs and Wonders
Christmas in Harmony
Life Goes On
A Change of Heart
The Christmas Scrapbook
Almost Friends

The Evolution of
FAITH

*How God Is Creating a Better
Christianity*

Philip Gulley

HarperOne
An Imprint of HarperCollins*Publishers*

HarperOne

HarperCollins books may be purchased for educational, business, or sales promo-
tional use. For information please write: Special Markets Department, Harper-
Collins Publishers, 10 East 53rd Street, New York, NY 10022.

HarperCollins website: http://www.harpercollins.com

HarperCollins®,■®, and HarperOne™ are
trademarks of HarperCollins Publishers.

FIRST EDITION

Library of Congress Cataloging-in-Publication Data is available.

ISBN 978–0–06–073660–6

11 12 13 14 15   RRD(H)   10 9 8 7 6 5 4 3 2 1

Contents

Foreword

In my final year of graduate work, I, along with my fellow students, was required to write a thesis describing my theology. Because the conferral of our degrees depended upon the quality of our dissertations, writing the theses was a matter of some anxiety for us. Stories of spectacular failures circulated around the school, embellished with tales of nervous breakdowns. One widely told story (I was never able to ascertain whether it was true) described the extended commitment of a student to a psychiatric hospital after receiving a failing grade. Despite my qualms, putting into words my understanding of God turned out to be a pleasant experience, not at all the onerous task I had anticipated.

Seventeen years later, as I was sorting through old papers, I happened upon a copy of my thesis and read it. While my life's experience confirmed some of my previous observations, much of what I had written years before made little sense now. Assertions about the character and activity of God, prayer, the purpose of the church, the nature of sin, knowing God's will, the person of Jesus, and the afterlife now seemed implausible, if not impossible. I could no longer affirm what I had once believed.

I wasn't saddened by this realization, nor did I have any desire to return to my former beliefs. Indeed, I felt energized by my theological journey, believing it revealed a vitality and passion often lacking in my more orthodox days. I had pitched my tent in a new, yet unexplored land, was pleased to be there, and wanted to investigate it further and map that territory for others by writing books such as this one.

Even as I reflected on the theological evolution in my life, I was also conscious of the language of my thesis, noting it was incomprehensible to anyone who hadn't studied theology. I obviously wanted to show my professors I was well-versed in theological jargon. Of course, one purpose of higher education is to teach a student the specific language of a given field. Medical students learn the language of anatomy and medicine. Law students become conversant in legal terminology. Those of us who study God learn the language of theology. But unlike the fields of medicine and law, people who haven't formally studied theology feel not only perfectly equipped, but duty bound, to engage in religious discourse. This is a compliment to our vocation, that those not schooled in the field wish to engage it. Unfortunately, we've not always made their participation easy, using language they neither know nor speak. Consider one of the more influential theologians of the past hundred years, Karl Barth, whose signature work, *Church Dogmatics,* was thirteen volumes containing over six million words. He worked on it for thirty-six years, and when he died still wasn't done. Many words can be said about Barth's work, but *accessible* isn't one of them. One has to wonder whether some who work in theology erect such barriers for the express purpose of excluding others, preferring the rarified air of theological speculation over a helpful, accessible spirituality.

The harm this causes is obvious—by excluding so many people from theological exploration, we increase the theological ignorance in our society, making people especially vulnerable to bad theology and unscrupulous purveyors of self-serving religion. A woman who lost her young child to leukemia was told it was because of unconfessed sin in her life. Because the woman was theologically uneducated, and the person pronouncing judgment buttressed his declaration with scriptures, she believed the pronouncement to be true and plunged into an abyss of guilt where she suffered for many years before seeking counseling. Others give money they can't spare to unprincipled hucksters who promise God's wealth and favor in return. Still others persist in abusive relationships believing it is God's will. On a far too regular basis, millions of people suffer the venomous effects of ill-conceived religion. That so relatively few people test theological assertions with reason, science, and logic is a testament to religion's ability to both seduce and silence its adherents. It is no wonder more and more thoughtful people are finding atheism an attractive choice.

In recent years, books written by atheists Richard Dawkins, Sam Harris, and Christopher Hitchens have earned a large and enthusiastic audience. I have found their books to be helpful and even compelling. Indeed, if push came to shove, I would choose their ethical humanism over many brands of Christianity I've encountered. But I would like to think that a well-reasoned theology, informed by common-sense, human experience, and the more enlightened aspects of religious tradition, can provide a helpful way forward for those of us who seek transcendent meaning, spiritual community, and joy.

For too many years Christianity has been more about constraint and less about liberation. Bound by dated creeds, traditions, and doctrines, the Christianity of the past has held too many of its disciples back, not carried them forward. This book is an invitation for you to consider your faith in a fresh way, informed by common sense, positive tradition, and personal experience. It represents, in a very real sense, an evolution of Christian faith. It is not an exhaustive theology. I write and think in broad strokes, preferring the sweeping beauty of a spiritual landscape over dogmatic minutiae. Nevertheless, I hope this book is a starting place, helping you think more deeply about the themes that might constitute an evolving Christianity for you. I write with no other agenda than to help your life become meaningful, abundant, and

joyful. Whether this book inspires you to become a Christian
or an ethical humanist is secondary to me. My dream for you
is a life well-lived, a life of compassion, grace, and dignity,
seeking and celebrating authenticity and integrity wherever
you find it. To that end, I invite you to read this book carefully, discuss it with your friends, and consider the questions
for reflection, trusting that a sincere examination of these
issues will bless your life, as it has mine.

The Evolution of Faith

Several years ago, at the invitation of a friend, I attended his childhood church on the Sunday it celebrated its one-hundredth anniversary. The pastor, in an exuberant moment, spoke of the enduring proclamation of the church, how since the time of Jesus the unchanging Christian gospel had been proclaimed throughout the world. The congregation nodded in agreement, even affirming their assent with robust "Amens." As a student of church history, I knew the pastor's claims were inaccurate, that over the past two thousand years, the church's message had undergone significant change, influenced by pivotal figures and movements. I even suspected that this specific church had experienced considerable theological change over its hundred years, reflecting the varied perspectives of its leaders.

I thought of the diverse mutations of Christianity I had encountered in my life—the Roman Catholicism of my mother, the Baptist leanings of my father's family, the Church of Christ tradition of a brother and sister, the Methodist perspective of another brother, the Presbyterian community of yet another sibling, and my own Quaker tradition. Each of those expressions emphasized a particular facet of the Christian experience. Each understood the mission of the church differently, employed differing styles of worship, and did not agree on how God could be known. They were not in harmony about the priorities of Jesus, and they did not share a common understanding of what it meant to be Christian, what it meant to please God, or how the church should be governed. Though they all bore the Christian name, their differences in belief were so considerable that one could reasonably conclude they represented different religions altogether. And that was just Christianity in the Western world. Were we to have stirred into the mix the many strains of Eastern Orthodoxy, the differences would have been staggering indeed.

Though I disagreed with the pastor's claim of an unchanged church, I understood his motives for making such an assertion. The church's authority rests on its declaration of doctrinal purity and her ageless, unchanging truths. Pastors who acknowledge the church's changing truth must convince congregants he or she nevertheless speaks with authority, a hard enough task in a culture already suspicious of institutional power.

At first glance, the title, *The Evolution of Faith: How God Is Creating a Better Christianity,* may seem presumptuous and egotistical, as if God were using me to liberate Christianity from its ancient moorings and carry it forward. But on a closer look, it makes perfect sense that if there are many versions of Christianity, that if Christianity has mutated and evolved over the centuries, it's reasonable to conclude it will continue to do so. It is also reasonable to conclude God might inspire a number of people to shepherd that process, that I might be one of them, just as you might be, and that a fitting response is to share our insights with others. Therefore, to speak of an evolving Christianity isn't to undertake a radical and unilateral overhaul of the faith, but to suggest a possible way forward that not only honors the ethos of Jesus but is conversant with our time and culture. For while it is clear that Christianity has changed and will continue to do so, what is less clear are the forms it might take.

What are the cultural factors that make an evolving Christianity inevitable? At the Quaker meeting I pastor, a woman of the Baha'i faith joins her Quaker husband in meeting for worship. Another attendee, a Jewish man, teaches an adult Sunday-school class; a young man in the congregation met a woman of the Muslim faith while in college; they married and are warmly welcomed into our meeting. Another man, intelligent and deeply caring, speaks openly about his leanings toward atheism. Imagine my standing at the pulpit and pronouncing these people spiritually lost, urging them to accept Jesus as their Savior. Not only would my sense of decorum prohibit that, but so would my appreciation for their obvious virtues. They are, to a person, loving, gracious, and wise. For me to suggest they are spiritually inferior would be not only unkind, but untrue.

At one time I thought such diversity was rare, but as I speak with my colleagues in ministry, I've discerned that more and more people are marrying outside their childhood faith, that those couples find joy and meaning in other spiritual expressions, and that many churches have incorporated these persons into their fellowships with sensitivity and warmth. This widespread openness to diverse religious traditions points to an evolving Christianity more tolerant than its predecessors.

In addition to this spiritual diversity, the pervasive acceptance of scientific advancements has significantly altered Christianity, especially those kinds of Christianity predicated on an outdated worldview. It is no longer possible for people to reject the scientific evidence of evolution without seeming ignorant. Nor is it possible, given what we know about homosexuality, to sustain a Christianity that asserts one's sexual orientation is chosen or inherently sinful. Add to this the stunning social changes brought about by the Internet, making spiritual and cultural isolation nearly impossible. Narrow religions can only be sustained when and where information is limited and controlled, when people are able to be "not of the world."

Case in point: at the urging of my publisher, I began a Facebook page. I have several thousand Facebook "friends" from a variety of geographical and religious backgrounds. Nearly every week, I post a theological or spiritual question, inviting responses. Given the diversity of my friends, the answers widely differ. I expected this. What I did not expect was the extent to which my Facebook friends would engage one another. Almost without exception, those exchanges have been cordial and sincere, with persons expressing much interest in the views of others, saying such things as, "I see your point. It makes a lot of sense. I'll have to rethink this."

The church's monopoly on Christian instruction is over. People feel quite free to join in theological discourse without the buffer of the church or its clergy. Were I in a religious tradition that emphasized the supremacy of a professional religious hierarchy, I would worry for my job, for it is apparent that more and more people are looking to spiritual resources beyond the conventional ones offered by the church. Whether people turn to a Facebook friend, or a TV talk show, or a neighbor, or a bestselling book, they are seeking religious counsel and spiritual insight outside the church. As they do, the church's authority, already weakened by scandal, abuse, and irrelevancy, will, I believe, evaporate altogether.

These cultural factors—religious diversity, advancements in science, expanded communication, and the church's diminishing role as the sole religious authority—are making the next stage of Christianity not only possible, but inevitable. Ironically, the more the church resists this evolution, the more likely it will hasten the change, for its efforts to preserve the

status quo will only emphasize its more negative strategies of
rigidity, control, and fear, thereby alienating the very people
it wishes to influence.

An Evolving Christianity Requires
an Emerging Theology

There has never been a significant shift in the church's
structure that wasn't accompanied by or inspired by a theological change. When Martin Luther initiated the Protestant
Reformation, he jettisoned many elements of the prevailing
theology, including the means of salvation, the authority of
the pope, and the necessity of priestly intervention for the forgiveness of sins. Whether a changing church is inspired by an
emerging theology, or a new theology materializes as a consequence of changes in the church, one is never seen without
the other. For change in the church never happens unless we
have convinced ourselves that God prefers that change and
thus have constructed a theology that justifies the changes
we've made.

I am no different. In my case, my experience of the
Divine Presence called into question many of the church's
practices, particularly the issues of institutional governance,
doctrinal authority, the scope of salvation, and the power of
grace. I have spent a good deal of my adult life constructing a theology that rationally supports the spiritual values I
first embraced from instinct. Some Christians have reacted

strongly to this, accusing me of heresy. What they fail to realize is how their own views, now considered traditional and orthodox, were at one time deemed revolutionary, if not heretical. A hundred years ago, their Christianity was the new Christianity.

A Preview of a Future Christianity

The theology in which many of us were raised fit hand in glove with the prevailing understanding of the church. It was exclusive, rarely acknowledging the merits of other religions. It emphasized a God above and beyond us, mirroring the ecclesial structure of the day that elevated leadership and concentrated power in the hands of an exalted few. It was decidedly privileged in nature and view, reflecting the cultural mores of the richest nations. Its God took their side, blessed their priorities, and helped secure their wealth and status. When Leonardo Boff, a Brazilian priest, criticized the church's alliance with the wealthy and powerful, he was accused of Marxism and silenced by the Roman Catholic Church's Congregation for the Doctrine of the Faith, led by Cardinal Joseph Ratzinger, who would later become Pope Benedict XVI. Under pressure from the Vatican, Boff eventually surrendered his priestly orders. This was all too common in a Christianity of dominance and control, but it will not stand in the emerging Christianity, whose philosophy will focus less on power and self-preservation and more on ecclesial modesty and service. Perhaps the evolved Christianity will ironically go back as it moves forward and will more accurately reflect the servant spirit of Jesus of Nazareth and be less concerned with worshipping Christ the King. For where the primary focus of a spiritual community is the worship of its central figure, the patterns of hierarchy become established, formalized, and perpetuated, eventually demanding unthinking conformity and unquestioned obedience.

My hope is that an evolving Christianity will reflect the egalitarian spirit of Jesus, not the elitism of an entrenched church. It will no longer presume that having male genitalia uniquely equips someone for leadership. Nor will it assume heterosexuals are capable of ministry in a way homosexuals are not. It will listen carefully to its young people, letting their enthusiasm and yearning for authenticity inspire a passionate and relevant faith. It will console the brokenhearted, speak for the voiceless, befriend the weak, challenge the powerful, and call to leadership those who handle power well—not for selfish gain but for selfless service.

An evolved Christianity will not insist we believe the absurd, affirm the incredible, or support a theology that degrades humanity. It will be a friend of science, working joyfully alongside the best minds in the world on a common mission to embrace and enhance life. This Christianity will talk less and act more. I recently attended a church gathering in which a committee had been asked to draft a resolution against torture. They had spent an entire year writing a short paragraph on which everyone on the committee could finally agree but no one else would likely read. When a woman rose to suggest they actually do something to prevent torture rather than just write words against it, she was criticized for not cooperating. People no longer listen to the church's pronouncements. No one waits with bated breath for the church to wade in with its perspective. We craft missives, epistles, and minutes that are first ignored, then forgotten. Nor do governments change their policies because Christians have collected on a street corner to sing "We Shall Overcome." But when ministers are bold and prophetic, when Christians rise from their pews and work and sweat and invest their lives, people take notice and lives are changed.

The richness of an evolved Christianity won't lie in slavish obedience to antiquated claims but in a vigorous commitment to care for the marginalized and an honest search for meaning and truth, no matter where it might lead. It is exciting beyond words to stand on the threshold of such a movement, to watch it unfold and flower, to watch it not only restore the church—which it just might, though that is not its purpose—but refresh and restore our world.

In the chapters ahead, I'll use as my framework the traditional areas of concern for Christian theology. Though that conventional structure is still an appropriate one, it is long past time its assertions were reexamined and reinterpreted in light of our changing world and expanding consciousness.

Perhaps you have not been accustomed to viewing faith from the vantage point of these topics, believing such matters are best left to theologians. But I believe these subjects have a tremendous influence on our personal spiritual journeys, helping us negotiate and navigate a more meaningful life. Just as the prophet Ezekiel saw a valley of dry bones stirring to life, so, too, can new life be breathed into our moribund faith, and God might say to us, as God said to those bones, "I will put my Spirit in you and you will live" (Ezek. 37:14 NIV).

Revelation: On Knowing God

What do the Islamic men who flew airplanes into the World Trade Center and the Pentagon on September 11, 2001, have in common with the Franciscan priest, Father Mychal Judge, who rushed to the World Trade Center that tragic day to help the dying? What do the gay Episcopal bishop, Gene Robinson, and Pastor Fred Phelps, the creator of a website called godhatesfags.com, both believe? What do the Catholic priest, the Pentecostal evangelist, the Jewish rabbi, the Islamic imam, the Quaker mystic, the televangelist, the Desana shaman, the Oxford theologian, and the elderly Nazarene woman volunteering on the telephone prayer line utterly and sincerely believe? What universal conviction is shared by almost every person who has ever embraced any faith?

Each is persuaded God speaks to him or her. Each believes that God communicated something he or she wouldn't otherwise know, through a book, a prophet, a pastor, a sign, an audible voice, an event, a religious tradition, a sacrament, a dream, or any number of means, except for God's intervention or revelation. Those who love deeply and those who hate deeply credit God for their impulses and opinions. When their beliefs collide with another's, they believe the other is mistaken, if not heretical or evil. The chief difference in their conviction is one of degree—the televangelist will say God speaks to him in an audible voice, while the Quaker mystic talks of leadings or nudgings. But each believes God has communicated something of the divine nature and purpose to him or her.

The root of all religious turmoil, which itself is the root of most earthly turmoil, is our vast disagreement over what God has said, how God has said it, and to whom. It is no exaggeration to say that millions of people have been murdered as a direct consequence of violent disputes over divine revelation. For sadly, divine revelation, by its very nature, deals with matters of such ultimate importance to so many, they would prefer death and perceived truth over life and perceived error.

I first became aware that others might have a contrary opinion of God's will when I was nine years old and moved with my family across the street from a Pentecostal family. My family believed God spoke through the Roman Catholic Church. Our neighbors believed God spoke through the Bible, as interpreted by their preacher. We believed God wanted us to abstain from meat on Fridays, confess our sins on Saturdays, and attend Mass on Sundays. They believed

God wanted them to attend church on Sundays (both morning and evening), Wednesdays, and Fridays. Each family was convinced the other was mistaken and that the penalty for their error was eternal separation from God. As children, we paid little attention to our theological differences, but when we became teenagers, our disparities became a topic of discussion, and finally debate. Had our neighbors not moved, I suspect we young people might even have come to blows over our differences, that we would have carried out on a small scale what has happened the world over on a larger scale—hatred and violence in the name of God.

Now, in the early years of the twenty-first century, these conclusions seem all too obvious: 1) There is little agreement about how God is known to us and scant evidence God dependably communicates with us, at least in a manner universally agreed upon; and 2) Despite this paucity of communication, we will fight and kill to defend our version of God's truth.

What Evidence Do We Have That God Speaks to Us?

On the shelves of my study is a collection of books given me by others. The volumes are of varying sizes and content but have one thing in common—each purports to be divinely inspired. There is a Koran, given me by the members of the first youth group I pastored in 1984, who believed it behooved a minister of the gospel to be acquainted with other religious traditions. Muslims believe the Koran was revealed to Mohammed by an angel named Gabriel. (Angels, it turns out, figure prominently in divine revelation. I would be more open to the existence of angels if I, or anyone I knew, had ever seen one.) Over the course of twenty-three years, ending in 632 CE, the angel Gabriel carefully dictated the Koran to Mohammed and his companions. As with all revelations, one has to take Mohammed's word on this.

There are Bibles on my shelf, the earliest one printed in 1832, the newest version a green Bible with a cover made of hemp (God's words about the environment are printed in green soybean-based ink). The Bible is not one book, but a collection of smaller books written over hundreds of years, taking its final shape in the fourth century CE. When Martin Luther wrote his own German-language version of the Bible, God told him to leave out four of the books—Hebrews, James, Jude, and Revelation—but God told other people to let them remain, so Luther compromised by placing them at the tail end of his Bible. Some Christians believe every word in these Bibles came directly from God and is literally true. Other Christians do not. Even within the same religion there is vast disagreement about how God has spoken.

Alongside my Bibles is a copy of *Conversations with God,* written not by Neale Donald Walsch, but by God, who, when Walsch was done writing a series of questions to God, took control of Walsch's pen and wrote the answers, or so

Walsch claims on the first page of his book. No one but Walsch was present when this happened, but he argues this point strongly and has convinced a legion of followers.

Several years ago, I was given a copy of *The Urantia Book* by a friend. It is 1,814 pages, and I haven't had the energy to tackle it. The book appeared, rather mysteriously, in Chicago in the second quarter of the twentieth century. Its authorship is unclear, though many of its followers believe angels and messengers from God had a hand in it.

The *Book of Mormon* sits next to my Urantia book. Joseph Smith is credited with writing the *Book of Mormon,* copying the text from golden tablets he discovered, with the help of an angel named Moroni, in 1823 in upstate New York. The golden tablets were presumably from God, but since Smith lost the golden tablets, we can't be sure. To be fair, Smith claims he returned the tablets to the angel Moroni. Moroni is not available for comment.

Often, when I speak somewhere, someone will ask if I have read *A Course in Miracles*. Though aware of the book, I've not read it, except to know it was written by Helen Schucman, with the assistance of William Thetford, purportedly under the direction of an inner voice over the course of several years.

A well-thumbed edition of the Western Yearly Meeting's 2005 *Faith and Practice* resides on my shelf. This book is from my own religious tradition—the Quakers. Compiled over a span of 150 years, it was initially proposed as a guide for daily living, not a creed. In time, as is often the case with religious writings, its contents took on divine authority—God-given, if you will—and any deviation from its declarations is now viewed by some Quakers as heretical. How a book that has been altered numerous times over the decades can now be viewed as God's unchanging truth is a mystery to me, but I know people who believe that to be true.

While there is no doubt value in each of these books, one could reasonably question how a divine being could dictate or inspire such a diverse collection of literature. Some have suggested this demonstrates God's commitment to diversity, but one could just as naturally wonder why, if God desired to reveal something of the divine nature, God didn't do it in a consistent manner, in a method understood by all people everywhere, especially given the significant harm caused by conflicting views of divine revelation. Why wouldn't an all-powerful, all-knowing, all-loving* God have anticipated the need for, and supplied the means for, a clear, consistent, irrefutable revelation? What is lacking on my bookshelves is the one text all agree is divinely inspired, for no such text exists, nor is it likely to. Indeed, even as global communication shrinks our world, what is expanding at an even faster clip are the religious and spiritual movements that claim unique revelations from God.

One response to diverse revelation is to assert that because

---

* While I am unconvinced of these attributes customarily assigned to God, the people most adamant that God has revealed something to them often believe God possesses these traits as well.

there is no shared consensus, God must not relate to us at all and any suggestion that God does is wishful thinking. Many people, I suspect an increasing number, believe this to be the case, hence the growing attraction of atheism. Weary of sectarian strife, they distrust any and all claims of divine leading. I would be hard pressed to refute their conclusions. Indeed, if God is the source of unity, as many religionists claim, it could be argued that the very division that often accompanies divine revelation is itself an indication of God's absence. For if God were authentically present in the ways we insist, it would result in unity and love, not the division and hatred that have often been its product.

Nevertheless, those of us who value a transcendent dimension to life hesitate to believe God is silent. Many of us, on some level, in ways we can't fully articulate or prove, have experienced God's presence (or what we believe to be God's presence) and in that experience have found direction, joy, and meaning for our lives. It also seems apparent that even those people who believe God has revealed the divine will in a manner different from that to which we are accustomed have also experienced that same direction, joy, and meaning, despite our claims of enjoying a unique revelation of God.

This lack of consensus suggests that the one thing we are most dogmatic about (divine revelation) should be the one thing about which we are least dogmatic.

Because we are most inflexible about that which cannot be empirically proven, we become defensive and unyielding, believing God is best served by dogged insistence rather than thoughtful seeking. There is a curious tendency I have seen repeated time and again: the more adamant we are about God, the less likely we are to embody the traits we believe God values—love, compassion, peace, wisdom, and patience. All of these virtues are forsaken in our efforts to refute the spiritual perspectives of another.

We are far too certain about something we have no proof exists. Where is the certain and obvious evidence God has spoken to us? Where is the irrefutable proof that God has revealed something of the divine nature to us? We have only the claims of others who've gone before us, who now are dead and can't be questioned. Where is the indisputable evidence that at a specific moment in time and history God appeared and conveyed a great truth? We are asked to accept those claims in faith, trusting in the witness of those we do not know, of those whose motives we're unsure of. Indeed, to take these claims on faith is seen as a virtue, as a sign of our love and loyalty to the Divine. To doubt these revelations is deemed sinful, an indication of blasphemy and worthy of punishment.

Why Do We Resort to Violence to Defend Our Version of God's Truth?

I once attended a gathering of Christians where a man stood to encourage those present to avoid biblical fundamentalism. He did so gently, warning against confining an eternal Spirit to the pages of a book. Later, while waiting in line for lunch, I overheard one man comment to another that the speaker should be beaten. I looked for any trace of humor, any indication the man was joking, but the anger on his face told me otherwise, that he truly believed a person should be punched and kicked for believing God might speak to us beyond the pages of Scripture. Incredibly, when we returned to our meeting, the very man who would have beaten his fellow Christian rose to speak about the dangers of atheism, claiming atheists lacked a moral foundation.

I left quickly after the event, not wanting to remain in such a toxic atmosphere. I wish now I had lingered and engaged the man in conversation in an effort to better understand his passions, if only in understanding him to avoid such extremes myself. Had we spoken, he likely would have told me it was his duty to defend God's truth, that the life of faith required vigorous resistance against all attacks. In that sense, his motives would have been no different than the Ayatollah Khomeini's, when he issued a fatwā, or death sentence, against the writer Salman Rushdie. They are all of a piece, the conviction that an unverifiable religious tenet is more precious than life itself. Armed with this conviction, all other virtues pale in value to the true believer. In their inverted spiritual economy, God requires malice not mercy.

This violence in defense of God's truth is repeated the world over. In the Judeo-Christian tradition from which I hail, only a few generations had passed before God demanded the brutal extermination of Israel's potential enemies. Of course, some would argue God did no such thing, that Israel used religious justification to defend its brutality, which only confirms my point—divine revelation, whether true or not, almost inevitably ends in violence.

Why is this?

Is it because the stakes seem so high, the consequences for blasphemy and error so severe, that we are driven to extremes in defense of divine truth? One has to wonder. In the Hebrew scriptures, those who misunderstood or angered God weren't gently corrected. They suffered terribly, losing their families, their land, and their lives. Revelation was a somber business, the consequences of misinterpretation high and harsh. Death was routinely meted out to those who got it wrong, under the convenient label of heresy.* Even today, those who fall outside the accepted worldview of their religion are still subject to harassment and exclusion, so desirous are some persons of faith to punish perceived nonbelievers.

How Is Revelation Best Handled?

My wife and I feed birds and enjoy sitting on our back porch watching the birds swarm around the food we set out for them. We especially enjoy hummingbirds and their aerial acrobatics. One evening, while winding down our day, we

---

* It is notoriously easy to be charged with heresy. Simply believe something a dominant religion doesn't and you are guilty of heresy.

watched as two male hummingbirds battled for dominance at the feeder. One began chasing the other, who flew beak-first into our garage door at such high speed that its needle-like beak penetrated the thin veneer of wood covering the door. The hummingbird, clearly stunned, found itself stuck and unable to free itself. I gently grasped its beak and pulled it free from the door, holding the hummingbird in my hand until it recovered.

I felt its heart race wildly, beating against my palm. Clutching it loosely, I knew I could easily kill it by closing my hand. Of course, I had no desire to harm it and simply studied the bird. It was a beautiful living thing, its throat iridescent and shimmery red, its body a jewellike green. Holding it, I felt as if I had been given a gift, an opportunity to enjoy what few others have experienced. I carefully passed the hummingbird to my wife who marveled at its delicate size, holding it loosely until it began to stir and eventually fly away. To this day, whenever we see a hummingbird, we turn to each other and say, often simultaneously, "Remember that hummingbird?"

I've often thought revelations and insights about God ought to be handled much the same way, loosely and softly so as not to smother or harm them. Unfortunately, this is usually the opposite of how divine truths are held. Our tendency is to grab them tightly, seizing them, squeezing out their vibrancy and vitality until life is gone from them. Indeed, one of the first things we do is codify and sanctify our encounters with the Divine. We try to recapture the moment by creating highly formalized sacraments in which these experiences can be relived. Recall the apostles' last meal with Jesus and how it has become ritualized in the sacrament of communion, becoming a standardized, often empty, routine. Of course, for some people it remains a meaningful moment, but I suspect for many more people it has become a rote event, no longer capturing both the joy and pathos of the original occasion. Our encounters with God often become similarly set in stone in our effort to sustain the freshness and power of the original insight.

When I was a young man, I had a revelation that I, and everyone else, was deeply cherished by God. It had a radical effect on my spiritual life, causing me to become a universalist. At the time, I was seated in my 1974 Volkswagen Beetle, parked outside the apartment I shared with my wife. Because that moment of divine insight was so beautiful, it would have been tempting to spend as much time in my car as possible, parked in that same location, hoping that one transcendent moment could be recaptured were I to replicate the setting. But I realized my experience had little to do with location and that transcendent moments can't be summoned at will.

Unfortunately, this is precisely what we tend to do whenever humanity and divinity intersect. We freeze the moment, believe it represents the totality of the divine character, insist that our encounter is superior to our neighbor's, and move quickly to define, and consequently limit, the manner in which God is encountered. If God spoke to us through the Bible, we believe the Bible is the sole means by which God communicates. If we meet God in nature, we return to nature again and again, believing that is the only place God is met. If we find God in the sacraments, we repeat the ritual each Sunday in an effort to find God once more. If God is discovered in the mosque or synagogue, we return to those places time and again. We do this because these transcendent moments tend to be beautiful and affirming. We often feel an infusion of new life and might even speak of being born again or enlightened or awakened or transformed or made whole.

Because our encounter with the sacred was so personally meaningful, when others tell us they met God differently, we question whether they met the true God. We wonder how their experience could be so different from our own, believing our encounter with the Divine to be the definitive one. In narrowing the means by which God is known, we diminish the God experience, turning to ice what was intended to be flowing water.

I had an elderly friend named Ray who never departed without sharing a word or two of advice. Sometimes, his counsel was humorous, other times serious. His favorite adage was, "Always entertain the possibility you might be mistaken." I would expand that by saying, "Always entertain the possibility your experience of the holy is limited." Or, as St. Paul famously declared, we know in part and see in a mirror dimly. Thus our revelations should be cherished, but loosely held, as one might hold a fragile bird, taking care that in our desire to protect it, we do not squeeze the life and power from it. Ironically, it has been my experience that the more loosely we hold such things, the more open we become to the Divine Presence meeting us and others, and the more likely we are to experience these sacred visitations. For the Spirit, like the wind, blows where it will, though seldom in our demanded directions.

How Do We Know Something Is from God?

For many years, my wife and I have not had television at home, so I look forward to an occasional stay in a hotel so I can indulge my fascination with religious broadcasting. Because I am theologically curious, I am intrigued by television preachers and their peculiarities. One televangelist has the odd habit of pausing while preaching, cocking his head as if listening to someone, then thanking God for speaking to him and sharing his revelation with his audience. Another televangelist reports regular conversations with God in which he is told how God would bless those who send the televangelist money.

The hoped-for effect of these "conversations with God" is that the hearer would find them compelling and believe God wants the hearer to do as the preacher asks. While many gullible people do, others become skeptical about every claim of divine leadings, even when those divine promptings might have been authentic. This is where we find ourselves today. Because revelations are often met with skepticism, some people are reluctant to share their leadings for fear of being thought odd. I recall an instance not long ago when a church was bogged down in a contentious disagreement about a sensitive matter. A woman in the congregation approached me and said God had shown her a way forward for the church while she was in prayer. When she shared her insight with me, I urged her to share it with others in the church, but she refused, fearing people might think her pretentious for offering words she believed were from God. "Besides," she said, "how can I be sure it was from God?"

Isn't this the question any thoughtful person asks at some time? How do I know an insight or leading I've received isn't a projection of my own wishes and will? What right do I have to claim my words are from God and should therefore carry more authority than someone else's?

How do we know when an insight or revelation is from God?

A woman I know was born into the Roman Catholic Church and continues to find it a helpful spiritual home. While discussing the topic of revelation with her, she told me God spoke through the traditions and hierarchy of her church. She believes God acts to preserve the church from error and that the teachings of her church and pope are infallible in matters of faith. Her conclusion assumes the church reliably interprets and conveys God's will, an assumption one is able to maintain only by not looking too closely at church history.

I have a friend who loves the Bible but, by his own embarrassed admission, hasn't read all of it. He grew up in a religious tradition that affirms the inerrancy of Scripture, finds that tradition meaningful, and embraces its claims of biblical inerrancy. When I put the above question to him, he answered, "If it is consistent with the Bible, it is from God." He is not the only person who has ever said this. Indeed, were we to poll a variety of Christians, we would hear this answer time and again.

His answer, of course, assumes there is an internal consistency to the Bible, that it speaks in a uniform voice that is readily discerned. But anyone who has read or studied the Bible in any depth knows that isn't the case. Written over a period of hundreds of years, by a myriad of authors, the Bible offers a variety of understandings about God and God's nature. In testing our leadings or revelations by the Bible, it is helpful to identify which voice or worldview in the Bible will be our benchmark.

These are just two examples within the Christian community of the differing tests of divine revelation. While each of them offers some insight into God's character and will, neither is wholly reliable. As one who believes Jesus revealed the values and priorities of God, I would test revelations by their consistency with the Jesus who welcomed the outsider, befriended the despised, loved his enemies, and called his friends to do the same. To be sure, there are other, less flattering, images of Jesus in the Gospels. At times, the Jesus of Scripture could be harsh, condemning those with whom he disagreed. Whether this is an accurate portrayal of Jesus isn't known. Still, in the interests of accuracy, it is important to acknowledge the multiple facets of the Jesus's nature presented in Scripture and to identify the specific Jesus image I use in my effort to discern which leadings and insights strike me as authentically divine.

To that end, I ask myself these questions: Will my obedience to this insight result in the betterment of others, especially the marginalized? Does this revelation call me beyond self-absorption and empower me to love and include others in my life? Is this prompting the most loving and gracious response I can imagine? If I can respond positively to these questions, I begin to suspect my insights might be divinely prompted. But because I can never be absolutely confident of that, I hold that revelation loosely, as one would a fragile living thing, not wanting to mistake my will for the divine will. But even as we realize the possibility of misinterpreting the divine will, this should never blind us to the presence of God's gracious activity in our world. We never know when we'll encounter the Divine Presence.

I have a friend named Lyman who's a retired school teacher. He is now deceased, but when I first met him, he'd just retired and in an effort to keep busy had begun serving the noon meal at a homeless shelter in the inner city of Indianapolis.

He'd been volunteering there a couple of weeks when a young man named Mike staggered in one day for lunch. The other workers at the shelter told Mike to leave, then explained to Lyman that Mike was a drunk and they didn't want him there. Mike continued to show up every day for lunch, and every day he was asked to leave.

Lyman didn't feel right about it, so one day when they threw Mike out, Lyman went out with him. Mike didn't smell like alcohol and didn't exhibit any of the other signs of alcoholism. He just staggered and fell into things and sometimes when he spoke it didn't make sense.

By then, Mike had been homeless for some time and didn't smell very pleasant. His hair was long, matted, and filthy. Lyman took him home, let Mike use his shower, took him to his barber and got him a haircut. It turned out that underneath all the hair and dirt, Mike was a handsome young man. Lyman and his wife, Harriet, fed him several good meals, let him sleep in a soft, warm bed, bought him new clothes, then took him to their doctor, who diagnosed Mike with Huntington's Disease, an incurable neurodegenerative genetic disorder that affects muscle coordination and some cognitive functions. People with Huntington's Disease stagger when they walk, and the disease destroys brain cells, which leads to dementia. If you didn't know someone had Huntington's, you might think the person were drunk.

Lyman lined up an apartment for Mike, arranged for a visiting nurse to check on him daily, and got him signed up for Social Security disability payments. Every day he'd go visit Mike, take him groceries, help him clean his apartment, and go with him on short walks around the neighborhood—this elderly man and this young man, the young man leaning into the older man. Eventually, when the disease got too bad, Lyman arranged for Mike to be moved to a nursing home, and when Mike passed away a few years later, Lyman was at his side.

We had Mike's memorial service at our Quaker meetinghouse. Mike didn't have any family living in the area, so it was just the Quakers, on a Sunday morning at the meetinghouse. Lyman gave the eulogy. The rest of us sat and listened as Lyman spoke about how he'd found the presence of God in Mike. Mike couldn't speak, of course, but if he could have, I suspect he might have talked about how he'd found the presence of God in Lyman.

This experience taught me two things about God and revelation. The first is God's habit of revealing himself in the broken and rejected. Those who've cultivated the habit of paying close attention to the marginalized often experience God's presence to a degree others have not. I would go so far as to say that compassion for the least of God's children, since it heightens our awareness of the Divine Presence, is the start of revelation.

The second lesson I learned was how our moments of divine visitation can never be planned. This is somewhat ironic, given that the church has spent much of its history carefully defining and constructing the means by which God is known. Instead, the Divine Presence makes herself known when we least expect, as we go about our daily tasks, entering our lives in the most unlikely disguises.

This, then, is how we know something is from God. It will call us beyond self-absorption, empowering us to love and include others in our lives. For God is love, and those who live in God live in love.

3
God

A few years after graduating from high school I enrolled in a variety of evening classes at a private college in Indianapolis. My first class was an introductory course in the Old Testament, or Hebrew Scriptures, taught by a man named Harold Tucker, whose name I remember for two reasons—he was my first college professor, and I fervently believed he was going to hell.

Prior to that class, my theological education had been rather sketchy. I'd been made to participate in catechism classes at a Catholic church, then had joined a Quaker youth group that emphasized activities and service over theological instruction. While my theological knowledge was limited, the few beliefs I did have were held with a deep intensity. I believed God was a male, was prone to violent and sudden outbursts, acted in ways we did not and could not understand, granted our requests if we prayed in faith, and was just as likely to condemn us to hell as send us to heaven. I further believed the Bible was the literal and inerrant Word of God and was to be accepted without question. Curiously, neither my parents nor youth group leaders taught me these things, but they were often articulated in the conservative, Midwest culture of my childhood, and I affirmed them. I'm not sure why I felt these beliefs to be true, but I suspect it had something to do with the adamant certainty of those who voiced them.

So when Harold Tucker strolled into class, flipped on the light switch, said, "Let there be light," then told us the first five books of the Bible were not authored by Moses, who had cleverly described his own funeral in the book of Deuteronomy, I suspected Harold Tucker was a heretic destined for hell. Then, when he said God was not male, I was sure of it.

I remember those early college years as a very painful time, when my understanding of God was laid waste and I was forced to reconsider everything I had presumed to be true about ultimate reality. Because it was a traumatic experience, I didn't fully appreciate the favor Harold Tucker was doing me and was angry with him for some time. While our childhood images of God are eventually tested and often found wanting, it is invariably bruising. Still, those who shepherd us through the death of inadequate gods are in the end doing us a favor, upsetting as it may be.

Years later, I had a conversation with a woman who had sailed for several days along the eastern seaboard of the United States, about fifty miles offshore. She was frightened the entire trip, feeling especially vulnerable because she was unable to see land and gain her bearings. The fear she described mirrored the anxiety I felt when my religious landmarks disappeared and I lost my spiritual orientation. I felt adrift, out of control, unable to chart my future. Gradually, my fears subsided, though they were intense enough that I still recall them and readily understand the qualms other people experience when their perceptions of God are challenged.

More recently, I attended a religious gathering where a person stood and said a book I had written had torn his life apart. His voice was tinged with fury, and he concluded his remarks by saying I was not qualified to serve as a pastor. I was later asked if I had been bothered by the man's remarks. I remembered my anger with Harold Tucker when his ideas had done the same thing to me and commented that I understood precisely what the man meant but shared my hope that one day the man might be grateful, as I am now grateful for the spiritual renovation Harold Tucker undertook on my behalf, though at the time I felt only splintered and broken apart.

While I didn't respond to the man, I've thought of that moment many times. Were it to happen again, I might encourage the man to understand the role of the pastor in a new way. For too long, the pastor's function has been that of propagandist, perpetuating a party-line view of God that is not always helpful or sound. When the pastor is the mouthpiece for a settled view of God and rewarded for his or her adherence to that view, the incentive to expand our understanding of God is lost, the church becomes spiritually stagnant, and the cause of truth is not well served.

The Real Blasphemy

As a child, I was told by a schoolmate that blaspheming the Holy Spirit was the unforgivable sin. I had no idea what that meant and asked him to explain his comment. He told me he wasn't sure what it meant but had heard his pastor talking about it and believed it had something to do with hating God, the penalty for which was immediate death.

I came home and told my neighbor, a boy named Mark, who suggested we test my schoolmate's theory. We hid behind a tree in Mark's backyard and simultaneously whispered, "I hate God," reasoning that if God did kill us and send us to hell, we would at least have the comfort of each other's company. Nothing happened. So we said it louder on the off chance God hadn't heard us. Still nothing. The birds still sang. I fell asleep that night and woke the next morning to a new day, very much alive.

At the time, I concluded my schoolmate knew nothing about God. But as I reflect on that experience now, I think more about the fear that undergirded his comment. Unfortunately, fear is a common emotion when people describe their relationship with God—fear they might believe the wrong thing and be punished, fear they might offend God's sense of holiness, fear God will condemn them to hell on some future day of judgment, fear their doubts will invite God's wrath, or fear they might inadvertently disobey God's Word.

In addition to fear, I have also seen its twin, anger, especially in those people whose view of God is trapped in the amber of childhood. Their beliefs hold up until they encounter difficulties that call into question what they've been taught. They discover God doesn't answer prayer the way they'd hoped, nor does God alter the rhythms of the natural world to meet their needs or expectations. The people they love are still visited with hardship, still get sick, and still die, despite the childhood assurances of God's active and specific care. To make matters worse, after discovering this chasm between reality and belief, they are discouraged from questioning God, reminded of God's power and wisdom, warned of our human fallibility, and finally urged to pray even more, that they might accept God's will with faith and trust. They often accept this recrimination, but underneath harbor a quiet resentment toward God for not holding up his* part of the bargain.

Fear and anger have hampered the spiritual lives of many people, binding them to an unhelpful view of God they need

---

* I know God is not a he. I also know God is not a she. But try writing a book about God without using personal pronouns. It's tough. I will use both pronouns, as it seems appropriate. Please don't write accusing me of insensitivity. Instead, devote that same energy to inventing and popularizing a gender-neutral, singular pronoun.

desperately to shed, lest their beliefs impede their emotional well-being and moral development. Of course, if one's relationship with God is primarily about indoctrination, fear and anger are bound to follow and "getting it right" becomes vitally important.

But what if exploration were the theme of one's spiritual journey? What if "rightness" were of secondary importance and what was paramount was the freedom to investigate unchartered spiritual ground? What if God were not honored by our commitment to orthodoxy, but by our willingness to traverse the difficult terrain of wisdom and discernment? If that were the case, God would not be owed our fear and submission, but our most probing questions. *True blasphemy would be ignoring our responsibility to engage the world and reality at the deepest level of which we are capable.* It would be to meet creation with apathy, with no appetite for inquiry, knowledge, or enlightenment.

In an address to the Philadelphia Yearly Meeting of the Religious Society of Friends (Quakers), writer Elise Boulding said,

> What is the Quaker faith? It is not a tidy package of words which you can capture at any given time and then repeat weekly at a worship service. It is an experience of discovery which starts the discoverer on a journey which is life-long. The discovery in itself is not uniquely a property of Quakerism. It is as old as Chris-

tianity, and considerably older if you share the belief that many have known Christ who have not known His name. What is unique to the Religious Society of Friends is its insistence that the discovery must be made by each man (woman) for him(her)self. No one is allowed to get it second-hand by accepting ready-made creeds. Furthermore, the discovery points a path and demands a journey and gives you the power to make the journey.*

Though Elise Boulding was speaking as a Christian and Quaker, her words apply to humanity's common search for meaning and our shared need to have an "experience of discovery." Healthy religion, and its leaders and teachers, should equip us to move beyond our stunted views of God so we are better suited for life and living. This is the difference between our *indoctrination* into a worldview that is no longer sustainable and our *exploration* of a new vision of God that is life-affirming, coherent, and beneficial.

Indoctrination vs. Exploration

Who among us has seen God? Who has taken God's picture or recorded God's voice? Who has offered verifiable proof of God's existence? While enlightened persons around the

---

* http://archive.pym.org/publish/fnp/Faith_n_Practice.txt

world have agreed on many matters—laws of science and nature and the principles of mathematics to name two— there is a vast lack of consensus about God. This has not kept us from making confident assertions about a God we have no hard evidence exists. The beauty of creation, the stunning capability of the human mind, and the complex arrangement of our physical world are often cited as confirmation of God's existence, but each of those wonders have scientific explanations more plausible than divine provenance.

This lack of consensus and proof should give us pause when talking about transcendent matters, but just the opposite is true—we are never as dogmatic as when we speak of God, despite our broad ignorance of the Divine. I say this not to be insensitive, but to point out the obvious—the fact that there is such wide disagreement about God can only mean that if some people are right about God, a far greater number are likely mistaken. Given this reality, shouldn't God be approached with less certainty and more open-mindedness? Shouldn't our approach to God be marked with a level of humility that reflects the vast gulf in our knowledge? But when the chief aim of religion is indoctrination, then humility, enlightenment, and open-mindedness fall by the way. Instead, efforts are made to "cement" our thinking early in life, encouraging us to accept the settled doctrines of the church. Traditionally, this has been done by urging children to either confirm their faith in more mainstream churches or to "accept Jesus" in more evangelical churches. Though the method is different, the goal is the same—to establish early in one's life a pattern of assent and obedience to religious beliefs the child can't yet possibly know to be true.

On a summer evening a few years ago, a family with whom we are friends joined us for dinner. After our meal, we moved to our porch to visit. One of the teenagers present began talking about his experience with organized religion. His parents belonged to a Christian denomination that emphasized the practice of confirmation for its children. At the age of twelve, the child was enrolled in confirmation classes, the culmination of which was standing in front of the entire congregation on a Sunday morning answering questions posed to him by the pastor. The boy had been made to memorize very specific answers to the questions, supplied by his confirmation teacher.

When the Sunday arrived for the boy to stand with the other children for confirmation, he told his teacher he had doubts about some of the answers he was expected to provide.

"Say them anyway," his teacher told him.

The boy felt uncomfortable affirming something he didn't believe but wasn't sure what to do. He took his place in line, marched into the sanctuary with the other children, and stood before the large congregation. The priest began asking them questions, working his way down the row of children, each of whom gave the predictable, memorized response.

The priest came to the boy and asked him the same question; the boy paused, then said, "Well, I'll tell you how I see it." He then proceeded to tell the pastor and congregation, in his own words, what he believed.

This was a first for the priest, who hesitated for a moment, was going to challenge the boy, apparently thought better of it, and went on to the other children, all of whom gave the approved response.

"Every time it was our son's turn, you could sense the pastor wanted to skip him but knew he couldn't," the boy's father said, chuckling. "And each time, our son said, 'Well, I'll tell you how I see it.'"

This is a courage rarely seen in organized religion, especially by one so young, in a religious culture that has emphasized indoctrination over exploration.

If the church has a future, it will lie in its ability to inspire its members to do what that young boy did—refuse to uncritically accept the settled answers of the past, resist the pressure to conform, and revere the continued search for meaning.

In practical terms, how would these changed priorities affect our understanding of God?

When I began studying the Bible in earnest, I was intrigued by how God seemed to change as a consequence of her encounters with humanity. The post-Flood God was penitent, vowing never to destroy the earth or living beings again (see Gen. 8:21). Angered by Israel's worship of the golden calf, God "repented of the evil which he thought to do to his people" (Exod. 32:14). It is unclear whether God changed or whether people created stories that reflected their evolving understanding of God. What is clear is that the idea of an adaptable God is not a new one, that God's adaptability almost always reflected God's movement toward a higher, more ethical state, and that the belief in an evolving God began early in the Judeo-Christian faith and is continuing still, despite the church's fondness for an unchanging and settled deity.

I was once leading a workshop for roughly one hundred pastors and asked them the question, "Do any of you still believe the exact same thing about God that you did when you were a child?" It is a question I frequently ask when speaking with pastors. On this occasion, the pastors chuckled and began talking about how their understanding of God had changed over the years. But one pastor raised his hand and said, "My beliefs about God have never changed."

Several of the other pastors approached me privately afterward and said, "He is telling the truth. He is the dullest, most unimaginative person you'd ever want to meet."

Though I didn't ask the man the motive for his spiritual stasis, I suspect he believed God was honored by his unchanging mind. I further suspect this man confused a moribund religiosity for steadfast faith.

But when you believe God is honored by a spirit of exploration, you begin to consider the possibility that God might be far more than we've ever imagined or even beyond our ability to imagine.

My younger son is fond of gadgetry and was once loaned a pair of night vision glasses by a family friend. We excitedly awaited the arrival of nighttime so we could explore the woods next to our home in the dark, aided by the glasses. While viewing the forest and meadows was interesting, what stunned us were the number of stars visible through the glasses. Because the glasses gathered and magnified all possible light, even the dimmest stars, normally invisible to the human eye, shone brightly. It was, in a word, staggering. In writing about it now, I struggle to find the words that can capture the experience. When I realize we were seeing only a tiny sliver of one galaxy in a universe of perhaps one hundred billion galaxies, and that some of the galaxies contain in excess of one trillion stars, I am left nearly speechless.

I feel the same way whenever someone asks me to describe, in a few short sentences, the infinite nature of God. I find myself resorting to describing what God is not, and doing that with some trepidation, realizing I've only seen one sliver of the God-experience in a world of billions of God-experiences. Nevertheless, if our understanding of God is to grow, it will happen because we've shared our insights, however small and tenuous, in hopes they might fill a vacant space in the human quest for understanding. That said, what follows are my current observations about God's nature, character, power, and will. While these descriptions are by no means exhaustive or conclusive, I offer them in hopes they might help you think more deeply and intentionally about God's presence and activity in our world.

The Nature of God

When I was a child, I would ask my mother questions about God. She would answer simply, in concepts and language I could understand. One day I asked her what God was like, and she gave the answer she had been taught. "God is all-loving, God is all-knowing, God is all-powerful." This answer satisfied me until I was around ten years old and saw a photograph of a young African American man chained to a tree and beaten. I had been told in church that God loves everyone and could do anything, so I couldn't figure out for the life of me why an all-powerful God hadn't intervened to help that man. This was the first crack in my spiritual foundation, the dawning realization that God might not be all I was taught.

In the next several years, other cracks developed as I learned about the Holocaust, war, disease, famine, ethnic cleansings, and other evils that multiplied human suffering and grief. Persons who clung to a belief in God's power offered various explanations for the presence of evil.

"Evil exists," they said, "because God has given humans free will and sometimes humans choose to do evil." While that handily explained war, murder, and other humanly initiated evil, it didn't explain tornadoes, tsunamis, earthquakes, floods, and other natural catastrophes that seemed to happen on a clockwork basis.

"But that is not evil," I was told. "That is the natural

rhythm of the world." While I knew that was true, I also knew those calamities were the cause of great suffering and something an all-loving and all-powerful God should have prevented.

"Evil exists," I was told, "because of Adam's sin." But why would a God who urged us to love and forgive others hold a grudge against all humanity for the sin of one man? Why would that same God be so resistant to forgiveness and reconciliation? How could a God who was all-loving be so doggedly determined to punish us far out of proportion to Adam's transgression?

"Evil exists," I was told, "because of our personal sin." But what sin did a child born with spina bifida commit? What sin did a shaken, brain-damaged baby do to merit the evil she suffered at the hands of her mother's angry boyfriend?

"Evil exists," I was told, "to teach us lessons we might not otherwise learn." But why would a God who is purportedly all-knowing be so limited in imagination that the only teaching tools available to him were evil and pain?"

Sometimes people of faith would admit to ignorance* and say, "We can't know now why evil exists, but one day we will know." This answer assumes that the reason, once known, will make sense of our suffering and somehow redeem it. There were, conservatively estimating, sixty million persons killed in World War II. What possible reason might an in-

---

* Not that admitting to ignorance is a bad thing. Indeed, those of us interested in religion should probably do it more often.

scrutable God offer that could make sense of such horrific evil?

No one ever said what seemed obvious to me—God not only lacks the power to command the physical world, God also appears unable to control human behavior. Many people are so reluctant to affirm that observation that they would rather worship a God who seems cruelly indifferent to human suffering than a God who is unable to control events to the extent they would like. So deep is their need for God to control the world, they would affirm the power of God at the expense of God's love, preferring a deity of unbridled might over a deity of infinite grace.

While I no longer believe God foresees and controls all things, I do believe God is at all times, in all circumstances, with all people, staunchly committed to the way of love. I would go further and say God is not only committed to the way of love, God *is* love. *God is that universal and inward impulse that inspires us to seek the best for others, to seek the growth of the beloved, which is to say everyone.*

The Universality of God

Around the time I entered first grade, I became eligible to participate in catechism classes at the Catholic church my family attended. Apart from the casual and sporadic conversations I'd had with my parents about God, this was the first time I was given explicit instruction about God. Over the next several years, I was taught many things about God, very few of which I still believe. While Vatican II had ushered in an era of openness and ecumenism, the effects of that council had not yet reached the shores of my hometown.

In that church, I learned the only way to God was through the Roman Catholic Church and its sacraments. I believed that for several years, until I began attending an evangelical Quaker meeting, when I then believed the only way to God was through a faith in the atoning blood of Jesus. I held to that view for some time, despite some misgivings I was afraid to voice aloud. In my late twenties, it dawned on me that the days of killing heretics were past and that I had a duty to speak freely and honestly about my emerging convictions, which were more universal and progressive than I had been taught.

Though I have rejected the salvific exclusivity of the Roman and evangelical churches, I do not dispute that there is but one way to follow God—the way of compassion, mercy, and love. Wherever those virtues are practiced, God is present, with no respect or regard for the religious boundaries we humans have devised. This is the sole test of godly religion: does this religion increase our capacity and ability to love? Whether God is called Elohim or Allah, whether the worship of God is centered in mosque, temple, shrine, or church, whether Jesus is honored as savior, prophet, or teacher, whether none or all of the religious dogma we value are met, if love is present, God is there.

I would go further and say that were someone to deny God, God's commitment to be lovingly present in that person's life would in no way be diminished. Unappreciated perhaps, but in no way reduced. For nothing can separate us from the love of God. So determined is God to love, it might be the case that even our refusal to love inspires God to love all the more. For love has a way of multiplying when and where it is needed most. Just as the broken child attracts our deepest sympathy, so it is that those who are most dispirited (*not-spirited or spiritless*) exert a pull on the mercy of God.

Intimacy with God begins when we first imagine God's presence in those persons most unlike us. On the occasion of an interreligious Thanksgiving service, I had the opportunity to hear the Sikh artist, K. P. Singh, deliver a brief message of the universality of God, in which he said, "If we can not see God in all, we can not see God at all." The reaction of the crowd was palpable. While ecumenical events tend to attract broad-minded people, K. P. Singh's ability to condense in a few words this profound insight had a startling effect. Around the room, people nodded in affirmation. Of course, this would be the case! Of course, any God meriting our wonder would necessarily, and gleefully, transcend our human distinctions. If we can not see God in all, we can not see God at all.

The Power of God

I mentioned above my belief that God not only lacks the power to command the physical world, but that God also appears unable to control human behavior. By this I mean it appears God lacks the overt power to intervene in time and space to alter events and circumstances. Whether this is intentional or unintentional, I do not know. This is not to say God is powerless. If God is that inward, persistent love, present in every life and situation, then we can rightly and boldly speak of love's power to inspire, ennoble, and transform those whom it encounters. This power is not a physical force exercised by an omnipotent God in response to our prayers and faith. There is no consistent evidence that God regularly and predictably wields that kind of power, despite our understandable wish that God did. It is the recognition that God's love, just as human love, has the power to ultimately transform life's circumstances, motivating us to act with a charity and courage we might not otherwise feel. I'm not even sure *power* is the correct word to describe God's influence, but I lack the vocabulary to more accurately describe the nuanced effect of God's grace, so will rely upon the following illustration.

Several years ago, my family and I visited the Grand Canyon. We stood on the edge of the South Rim and marveled at the vast expanse of the canyon and the Colorado River twisting its way through the floor of the valley a mile down. Though we knew the river had carved this canyon over a period of forty million years, and perhaps even longer, it was still difficult to imagine that the same water that falls so gently on the Midwestern fields of our home could have altered these layers of rocks so dramatically.

Can the persistent presence of love have a similarly dramatic effect on human lives, wearing down our hardness of heart? Yes, of course. I have seen if firsthand. Does it happen as quickly as we would like? Hardly ever. But there is progress, however glacial. There is a power present in this slow dripping of grace, which eventually cuts through the dross of hate, revealing the goodness beneath. Should boulders of hatred attempt to restrict this current of grace, then grace like flowing water will find another path, will over time erode the stone and freely flow. This is the power of God, exercised not in overwhelming and sudden force, but in gentle and gracious persistence that eventually, but inexorably, makes beautiful the landscape of our lives.

The Will of God

Now let us continue with this metaphor of God and grace as flowing water and talk about the will of God. To speak of the will of God is to assume God is determinedly committed toward a particular goal, be it reconciliation, holiness, perfection, spiritual wholeness, or any number of states or conditions religions have emphasized. To speak of the will of God is to say that God has a distinct intention. What is the purpose of love? Isn't it always the emotional, intellectual, spiritual, and relational growth of the beloved?* Love is unswervingly committed to the betterment and completion of others. This wholeness is the hope and goal of parents who love their children, of persons who love their spouses or partners, of teachers who love their students, of therapists who love their clients, of pastors who love their congregants, of friends who love one another, and of all who love themselves. Consequently, to say God is love is to say God wills the growth and wholeness of creation.

Over the years, persons have come to me expressing concern that they weren't following God's will for their lives, inferring God's will was a very specific and unique course of action God intended their lives to take. They believe God wants them to marry certain persons, work at specific vocations, have a certain number of children, live in a particular home, spend their money a certain way, and attend a specific church. They worry they have misunderstood God's particular will for their lives and consequently are living outside the range of God's blessing and protection. They believe God not only has a specific plan for their lives, they believe God has a unique plan or will for each person and that our earthly task is to discern and follow that single path in order to be happy and fulfilled.

---

* I am indebted to psychiatrist and writer Scott Peck for this understanding of the purpose and activity of love.

It is almost impossible to exaggerate the number of people who worry, obsessively so, that they are living outside God's will, causing themselves unnecessary emotional distress in their belief God will punish them for their life's direction. While I admire these persons' desire to be faithful, I'm confused by their anxieties. First, such a belief is contrary to our conviction that God is love. Why would a God of love be primarily about punishment and retribution? Secondly, it seems contradictory to speak of a God who has given us free will and at the same time has devised a precise plan for our lives that must be followed for us to be blessed. It suggests God is double-minded, extending the gift of human freedom one moment, then punishing us for exercising that freedom the next moment.

Every thoughtful parent understands well the relinquishment of power that necessarily accompanies the growth and maturation of his or her children. While we have dreams for our offspring no matter their ages, we also recognize the crippling effect of managing the minutiae of their lives, as we did when they were toddlers. At some point, generally during their high school years, our direction and guidance moves from the specific to the general. We entrust more decisions to their oversight, resist offering unsolicited advice, and generally encourage them to assume more responsibility with the goal of emancipation and self-sufficiency.

Do we always approve of our children's choices? Of course not, but we know our efforts to spare them the consequences of their decisions might well hamper their future ability to wisely decide and discern their life's path. So, too, God happily grants us the freedom to make our own decisions, knowing we won't always be wise, but that even our poorer decisions are ultimately instructive and perhaps even redemptive.

While there are limitations to the analogy of God as parent, in this sense it is helpful. Because love is committed to the growth of the beloved, God is necessarily committed to our moral and spiritual progress. Consequently, God's will for our lives can only be general, not specific, for specificity invariably involves the micromanagement of our lives, which ultimately diminishes our freedom, growth, and responsibility.

To speak of God's will is to speak of God's general hope for our lives, which is our growth in love and mercy, our increase in wisdom, and our commitment to justice and integrity. Thus, God's will is not a tightrope that must be traversed with no room for error. Seeking God's will does not mean anxiously discerning God's preference on every matter large and small, worrying constantly about a misstep and a plunge into the abyss. *Seeking God's will means giving careful attention to the ultimate priorities of God, which are love, mercy, wisdom, justice, and integrity.** It is to keep these priorities ever

---

* The virtues of love, mercy, wisdom, justice, and integrity are especially esteemed in the Christian tradition that has informed my understanding of God. I realize other traditions might attribute different virtues to God, such as hospitality, sacrifice, sexual abstinence, or even warfare. This incongruity reinforces my maxim about revelation: the lack of consensus (about divine revelation) suggests that the one thing we are most dogmatic about (divine revelation), should be the one thing about which we are least dogmatic.

before us, letting them inform our lives so that all we are and all we do serves to expand these virtues. It is to consider such questions as these:

- Will what I am about to do result in the growth and betterment of others?

- Will this action increase love or diminish it?

- Will humanity's wisdom be expanded by my efforts, or am I appealing to ignorance and narrow-mindedness?

This is all to say that God is a generalist, not a specialist. As a generalist, God is concerned with the broader direction of our lives, which is our commitment to the betterment and growth of the beloved. In this endeavor we are not without resources. The power of God, which is the tenacity of grace, comes to our aid, transforming the world, just as water reshapes stone. Trusting in grace, we needn't worry about indoctrination, either our own or others'. Instead, we are free to explore the spiritual landscape, probing and questioning, discovering new territory that might ultimately prove more bountiful and beautiful than the settled lands of tired traditions. If Christianity is to evolve, as it surely must if it is to thrive, we must first unchain ourselves from the weight of dead habit that has dulled our minds and stilled our spirits.

When God is understood as that universal and inward impulse that inspires us to seek the best for others, to seek the growth of the beloved, which is to say everyone, we are able to move from fear to freedom, knowing God wants only the best for us and never the worst. As Clarence Jordan, the writer and preacher, once observed, "God is not a celestial prison warden jangling the keys on a bunch of lifers—he's a shepherd seeking for sheep, a woman searching for coins, a father waiting for his son."*

---

* Clarence Jordan, *The Substance of Faith and Other Cotton Patch Sermons* (New York: Association Press, 1972), 152.

Jesus and Jesus-Types

When I was a child, I was asked to play the role of Jesus in a church play. I was proud to be asked, believing my selection had something to do with my personal virtue. But as the teacher handed me the costume, she said, "You're the only boy skinny enough to wear this." This in no way diminished my enthusiasm for the part, and I played my character with gusto—taunting the boy dressed as Satan, jeering the Pharisees, and condemning several of my foes to the fires of hell.

Later, the teacher asked the class what each of us might do if we were given the powers of Jesus. Though this was many years ago, I seem to remember the girls used their power toward more benevolent ends—healing the sick, securing world peace, and feeding the hungry, while we boys used our power to dispatch our enemies, enrich ourselves financially, and win at sports. Though I wasn't aware of it at the time, I eventually realized our human propensity for creating the kind of Jesus we would be, or the kind of Jesus whose priorities matched our own.

Years later, when I studied the history of Christianity, I saw this tendency confirmed. Jesus was on *our* side when we engaged in battle, he protected *our* nation, saved *our* souls, healed *our* sick, blessed *our* finances, forgave *our* sins, and even died for *us*. It should come as no surprise that the church interpreted the role and work of Jesus to be primarily about saving us, an ironic and self-centered interpretation of a man whose very life called us away from egoism. While many interpreted the work of Jesus in this self-interested manner, other people understood Jesus's mission in a broader, more universal context. Jesus was the one who loved the outsider, who saved the world, who called us to love our enemies, and do good to those who despised us. But whether one's theological perspective was narrow or universal, the virtues we ascribed to Jesus often reflected our own understanding of ultimate good so that our image of him mirrored our own priorities.

This is to say that if one values inclusion, forgiveness, and mercy, it is entirely possible to find Jesus stories that reinforce that kind of Jesus: the celebrated compassion of the good Samaritan, the tender pardon of the woman caught in adultery, the sympathetic healing of lepers, the table fellowship with Zacchaeus the tax collector, the ebullient grace of the forgiving father toward his prodigal son. It is an easy matter to characterize Jesus in this positive light, to cite Scripture that confirm these virtues, and to earnestly believe we have defined the totality of Jesus's personality.

It is just as simple, if one values exclusion, righteousness, and judgment, to find Jesus stories that support that image of Jesus: the Jesus who violently overturned the temple tables, the Jesus who came not to bring peace but a sword, the Jesus who judged his fellow Israelites to be vipers, hypocrites, and blind guides, the Jesus who would divide father against son, the Jesus who cast people into utter darkness and balked at healing a Gentile woman. It is an easy matter to characterize Jesus in this hard light, to cite Scripture that confirm these qualities, and to earnestly believe we have defined the totality of Jesus's personality.

I now find it helpful, when people begin speaking with me about Jesus, to ask them to clarify which Jesus they are talking about, lest they speak about one kind of Jesus while I have in mind another.

This theological split-mindedness is not confined to Christianity. When one listens to the range of voices across Islam, one could believe Mohammed was either merciful or brutal. Indeed, the emerging schism in Islam between moderates and fundamentalists is rooted in their conflicting images of Mohammed's priorities and his expectations for those who followed him. Though I suspect this phenomena is present in every religion, these are two of the more obvious examples of our tendency to create our spiritual heroes in our image.

Our tendency to make *our* priorities the priorities of God

conveniently excuses us from the hard work of considering God in a new light. *When God is the extrapolation of our highest principles, we are seldom challenged to expand our consciousness, which is why the divine in any culture seldom rises above that society's collective morality.* Expansive, generous societies have expansive, generous gods. Provincial, self-interested cultures have provincial, self-interested gods. *This is why prophets, why Jesus-types, are necessary—they help us see God in new ways by challenging our beliefs and customs, thereby raising our collective consciousness.* For example, recall how Jesus would precede his teachings by saying, "You have heard it said . . . , but I say to you . . . ," then offer an alternative vision of God his hearers might never have considered—to love their enemies, to not seek vengeance, to not kill or lust or make exaggerated claims of honesty. This was done to help people grow beyond their present definitions of righteousness. It seems evident that persons other than Jesus have also committed themselves to this important work.

Spirit Persons, Jesus-Types, or God-Bearers

As I studied world religions, I realized the central figures in many other religions played much the same role as Jesus. They embodied divine qualities, utilized their intimacy with God to bless and enlighten others, and helped their followers experience an expanded view of God. Again, these Jesus-types were not confined to any one religion, tradition, culture or era, which made me conclude that while I discovered the character of God through the life of Jesus, others had arrived at the same insights through the life and witness of others.

Writer and scholar Marcus Borg has called such persons "spirit persons." These are the people, present in every religious tradition, who embody divine wisdom, who show us the way, and expand our God-consciousness. They are noted for their compassion and their deep care for those who suffer. In some instances, they marshal divine power to bring healing to the sick and lame. They call us away from self-interest to a higher regard for others and teach us what it means to love. While they inhabit temporal time and space, they also seem to occupy another realm beyond and above time, enjoying a fuller spiritual dimension most of us experience only in snatches and glimpses.

While studying theology in college, I encountered the Greek word *theotokos*. Historically, it has been used to identify Mary, the mother of Jesus. Used primarily in the Eastern Orthodox tradition, the word *theotokos,* translated literally, means "God-bearer." In that context, it describes Mary's biological role in the birth of Jesus. When I first learned the word, I thought of others I've known who seemed to birth the presence of God. Now I wonder if "God-bearer" might also describe the spiritual pioneers or prophets who bear the divine values to their culture. Some would say we are all called to be God-bearers, but certain people are clearly more adept at that than others, and a few people are so marvelously suited and gifted for that mission that those who know them say, "Surely the Divine Presence abides in that person in a powerful way."

As a Quaker, as one who believes there is that of God in every person, it nevertheless seems apparent that some have said *Yes!* to the Divine Presence more fully than others. While my assent to God is often tentative and halfhearted, Jesus's embrace of the divine will was rich and full. But that degree of participation with the divine has not been limited to Jesus, for persons in different religions can readily identify others who've done the same—Confucius, Buddha, Moham-med, Bahá'u'lláh, and Gandhi, to name a few. Though the temptation is always strong to claim a unique status for our God-bearer, it appears other traditions have also been blessed with bearers of the divine or Jesus-types.

When I have mentioned the possibility of other Jesus-types, I've encountered resistance from some Christians who believed our faith in Jesus required us to dismiss the possibil-ity that God might have equipped others to bear the Divine Presence as Jesus had. However, the value of Jesus did not lay in his uniqueness, but in his devotion, compassion, and wisdom. The notion that God might have considered Jesus worth replicating should not threaten us. Rather, we should fervently hope God would call others to embody the Divine Presence as richly as Jesus did. Our world needs more God-bearers, not fewer.

Regrettably, we've not often welcomed God-bearers. They are met with persecution and violence, usually because they represent a threat to our assumptions, doctrines, and beliefs and thereby jeopardize our power and its expansion. This is an important characteristic of God-bearers—they challenge systems of domination and call us toward a more egalitarian, gracious, and mutual way of life. Because earthly power is often rooted in religion, which either bestows power or presumes to bless it, Jesus-types challenge our spiritual standing and consequently our hold on power, which we fight to retain. This is why the contemporaries of Jesus-types don't always welcome them and might even silence them, fearing their personal power and cultural privilege might be threatened.

I was once employed by a corporation that had been run for years by a network of entrenched and powerful leaders. When the top position in the personnel department opened, a man from outside the company was hired to fill it. The man was unusually perceptive and quickly sensed employee morale was low. He organized a series of meetings involving every company employee, asking how our workplace environment could be improved. Initially, the meetings were emotionally intense as long-time workers vented years of pent-up anger. But as the meetings progressed, the responses became more thoughtful and constructive. For the first time in memory, employee concerns were taken seriously by someone with power. Morale lifted and a new sense of hope infused the company. Then rumors began circulating that the man had upset the wrong people, and with startling speed he was fired and the remaining meetings canceled. It was nothing less than a corporate crucifixion.

This pattern has been repeated the world over, both inside and outside religious institutions. Whenever entrenched power is threatened, whether by Jesus-types or not, the reaction of the powerful is swift and severe. In spiritual communities, Jesus-types are often silenced by powerful religionists, suppressed by the very people they had hoped to redeem. In the Gospel of Luke there's a telling line spoken by a Roman centurion at the scene of Jesus's murder. Witnessing what had taken place, he said, "Certainly this man was innocent!" (23:47) It is a tragic fact of history that we don't recognize the value and integrity of God-bearers until it is too late, when time has granted the perspective and ability to look back and say, "Ah, there was something powerful and truthful in that person's words." Think how much richer our lives would be were we to recognize these God-bearers while they were alive, take fuller advantage of their presence, and heed their words.

How Do We Recognize God-Bearers?

On November 18, 1978, a man from my home state of Indiana, Jim Jones, made headlines around the world when he led his followers in a mass suicide in Jonestown, Guyana. Broadly respected as a civil-rights leader and pastor in Indianapolis and San Francisco, in the early 1970s Jones began preaching that he was the reincarnation of Jesus, attracting large numbers of people to his People's Temple in San Francisco. In 1977, Jones and his followers moved to the country of Guyana in South America where he and 917 of his followers drank Flavor Aid laced with cyanide and died.

The annals of humanity are populated with persons who claim divine origins, most all of them sincere, but also deeply troubled. During my college years, I worked at a mental-health agency where I routinely met persons with messianic aspirations. Jim Jones was one of the more notorious claimants, because his intelligence, political power, media prowess, and charisma gained him a significant following. But for every Jim Jones there are thousands of others who, while lacking his skills for self-promotion, are nevertheless persuaded of their greatness. This has understandably made us wary of self-proclaimed messiahs. Skepticism in religious matters is probably appropriate, protecting us from those who use religion for their own selfish ends. But too much skepticism can blind us to the authentic presence of God in people and situations. Somewhere between the extremes of the messianic complex and spiritual nihilism is found the balance that permits us to recognize and appreciate authentic God-bearers.

When I was purchasing my first car, my father went with me to visit car dealers. One salesman we spoke with made it a point to mention his honesty over and over again. I was pleased to hear of his integrity and was eager to buy a car from him, so I was confused when my father bid the man farewell and quickly left the dealership with me in tow. As we were driving away, I asked my father why we hadn't purchased a car from that man.

"If a man keeps telling you how honest he is, keep hold of your wallet," my father said.

Over the years, I have found my father's observation to be generally true. Whenever someone adamantly claims to possess a particular virtue, it is almost a certainty he or she does not. Those who repeatedly emphasize their Christian faith often act most unchristian. Those who boast of their wisdom are usually foolish. People who speak of their fidelity are often snared in the web of unfaithfulness. Those who try to convince the rest of us they enjoy an exclusive relationship with God are often spiritually corrupt. This is to say that people who tell us they are God-bearers probably aren't. *Those who bear the Divine Presence will not have to point it out to the rest of us. We will see it in their conduct, in their humility, in their grace to others, in their compassion for the hurting, in their truth-telling that helps us grow, in their challenge to evil, in their courage in the face of abusive power, and in their yearning for justice.*

In one of the first meetings I pastored, a man took every opportunity to share his beliefs. While some people are inordinately interested in theological discussion (me, for instance), every conversation with him ended with his veiled conclusion that his views and conduct were superior to everyone else's. It became clear he expected the rest of us to share his point of view or incur God's wrath and judgment.

It also became evident he thought God had appointed him to save our congregation from a perceived spiritual decline. The consequence of this was that disagreement with him became disagreement with God's will, so every thought contrary to his own was met with hostility. When pressed to explain the circumstances under which he had been "appointed by God," he became angry and defensive, demanding that others acknowledge the gifts and power he believed God had given him.

Another man in the congregation was his opposite— gentle, serene, helpful, and gracious. He quietly invested himself in the lives of others, encouraging and assisting them in whatever way he could. The pattern of his life was one of availability, reaching out to the hurting and lonely, joyfully making time for those who needed whatever support he could offer. While deeply interested in matters of faith, he never vociferously defended his beliefs and seemed quite willing to examine faiths other than his own, able to see the truth they might offer.

While I would hasten to affirm that God could work through either one of those men, it was clear the second man seemed more sensitive to the Spirit's whisper and had developed his capacity for divine usefulness in a way the other man had not.

This is another trait common to God-bearers or Jesus-types—it isn't that God has preordained them to serve a special function. They have simply developed their ability and willingness to respond to the Divine Presence to a degree that enables them to be particularly supportive of God's purposes and humanity's advancement. Again, it is not that God has uniquely called them, for God calls each of us to the work of spiritual sensitivity and human growth. What differentiates God-bearers from the rest of us is the degree of response. Their affirmation of God is every bit as dynamic and complete as God's affirmation of them. They are God-bearers not because God had determined, long before their births, that they would serve that role. *They are God-bearers because they permitted themselves to be enlivened by the Divine Presence to a degree most of us have not.*

Consequently, another distinguishing characteristic of God-bearers is their enthusiastic and unfeigned commitment to the priorities of God, coupled with a passion for human growth and advancement. Authentic God-bearers move us further along the path of spiritual and human evolution. In this sense, they are progressive, not regressive, revealing fresh insights, leading the way to unexplored spiritual territory and innovative ways of living that honor and esteem those we've overlooked. This is the test of authentic God-bearers, spirit people, or Jesus-types—their unity with God is evident to all who encounter them. Our response to the God-bearer is then a test for us—do we welcome the God-bearer with respect, or do we perceive the God-bearer to be a threat to our personal priorities and silence him or her?

How Do We Respond to God-Bearers?

Many years ago, I met a fellow minister whose church was near the meeting I pastored. He was a charismatic individual, his church was growing rapidly, and his reputation with it. Whenever our shared vocation brought us together, I noticed other pastors deferring to him, soliciting his opinion, and aligning themselves with him. Indeed, his charm and appeal were so compelling, I found myself doing the same things in an effort to gain his approval.

Occasionally, I would encounter persons in his church who would invariably speak of their admiration of him. In one instance, a lady confided in me that she sometimes wondered if he weren't the second coming of Jesus. The handful of people in his congregation who were leery of him were encouraged to worship elsewhere. Within a year, he had restructured the church's hierarchy, installing elders whom he could control. His salary increased dramatically and he was given a luxurious car by the church. His sermons, taped and widely distributed, contained glowing self-references, highlighting his faith, wisdom, and commitment to God. His birthdays and anniversaries became opportunities for elaborate parties in the church, where he was showered with gifts and tributes.

But after several years, rumors began to circulate about his infidelities with various women in the church. Attendance began to decline as more revelations came to light, and donations to the church plummeted. Though he insisted on his

innocence, enough doubt had been raised to make his continued employment untenable. Within a few weeks, he quietly
left the city with his wife and children.

Shortly after his departure, I bumped into one of his
church members, who in the past had been most supportive
of his pastor. "I guess he was just evil," the man told me.
"And we didn't see it."

I have thought of that pastor many times since then. While
what he did was certainly inappropriate, I'm not persuaded
he was the only one to blame. Perhaps anyone might have
become spiritually and morally unbalanced if given the adulation and power he had been granted. It was as if an unspo-
ken agreement had been reached between the minister and
his congregation—he would symbolize spiritual purity and
power in exchange for wealth and adoration. It worked for a
time but could not be sustained, resulting in ecclesial divorce.

That incident caused me to begin thinking seriously
about the dangers of adulation. Unlike that pastor, Jesus was
seldom the recipient of uncritical adoration, probably because
he never demanded it. His humility was a safeguard against
the sense of self-importance that sometimes accompanies
spiritual leadership. Authentic God-bearers are uncomfortable being worshipped, knowing the risk of arrogance for
themselves and heightened, unreasonable expectations of
those they lead. The hope of Jesus was not to be venerated,
but to be taken seriously, for his vision of God's reign to be
weighed, evaluated, and ultimately embraced. Instead, the

church soon made Jesus its primary focus of worship, to a degree inconsistent with one he ever requested or was ever offered by those who knew him best. There were two principal responses to Jesus: curiosity coupled with respect, or anger coupled with self-righteousness. These reactions are still typical whenever a God-bearer or Jesus-type presents us with a vision that challenges our suppositions about God and one another.

Throughout the Gospels we read stories of people seeking out Jesus, asking him to make sense of the difficulties they encountered. To be sure, some people wanted quick relief from the problems they faced. But others, sensing his uncommon wisdom, engaged him in conversation, seeking to better understand the complexities of life. In the ninth chapter of the Gospel of John, the disciples of Jesus, encountering a man born blind, ask Jesus, "Rabbi, who sinned, this man or his parents, that he was born blind?" Jesus said the man's blindness wasn't caused by anyone's sin, but was an opportunity to demonstrate God's healing power. Whether we agree with the idea that some people suffer so God's power can be revealed isn't the point. What is pertinent is that many people believed Jesus could speak to their deepest existential questions, respected his perspective, and sought it out, curious to know his thoughts.

Other people responded to Jesus, not with curiosity, but with anger, especially when his vision challenged their worldview or power. Ultimately, he was killed by political and religious leaders who rightly perceived that his teachings eroded their legitimacy and influence.

Because God-bearers often inspire religions or movements, it is not unusual for them to become objects of veneration. This is understandable, even if it were never the intent of the God-bearers. Their lives and works resonate with others, words of appreciation and tribute are offered, stories of their feats are recalled and written (and usually exaggerated), their exceptional insight affirmed, and the seeds of a new religion are sown.

Unfortunately, in elevating our God-bearer, we often think it necessary to denigrate or diminish other God-bearers, as if truth were a contest and ours can only be right if others are wrong. So Christians disparage Muslims, who vilify Hindus, who fear Buddhists, and this circle of suspicion goes on, reaching into every nation and culture around the globe. This, despite the fact that God-bearers have carried the shared message of love, compassion, tolerance, justice, peace, and care for the poor.

The keys to humanity's survival are not only our ability to sustain a viable physical environment, but our willingness to recognize God-bearers in addition to our own. A friend of mine believed he was called to ministry and began the necessary training to become a pastor. The process took several years, and at its conclusion he was invited to meet with an ordination board who would discern his suitability for ministry. A man on the board asked him, "Do you believe Jesus is the only way to God?"

My friend, aware that the denomination to which he belonged was riven by theological differences, knew his assent would be expected if he wished to be ordained. Nevertheless, he couldn't in good conscience affirm Jesus was the only path to God. Spying a map of Indiana on the wall, he asked the board, "How many roads lead to Indianapolis?"

They remained silent.

My friend stood and walked to the map. "I get to Indianapolis this way," he said, pointing to an interstate. "But I have friends who live here," he said, identifying a town far across the state, "and they get to Indianapolis this way," he said, pointing out another route.

This generosity of spirit is essential for the world's future. It in no way dishonors Jesus or any other God-bearer to recognize that God has generously provided many paths to the divine. *Indeed, it verifies their value by acknowledging that the gifts present in them merited duplication in others.* While the church has often cited the words found in the fourteenth chapter of the Gospel of John—"I am the way, and the truth, and the life; no one comes to the Father, but by me."—most reputable scholars believe those words didn't reveal Jesus's self-understanding as much as they revealed the early church's eventual understanding of Jesus. It is striking to consider how many God-bearers the church has dismissed over the years because of a literal interpretation of words Jesus never likely uttered.

How We Can Be Like Jesus
and Other God-Bearers

If God-bearers are those who've permitted themselves to be enlivened by the Divine Presence to an extent others of us have not, it stands to reason that the only thing keeping us from doing so is our own unwillingness or inability to engage the Divine Presence at a richer, deeper level. I'm presuming God desires this fuller relationship with all her children and has not singled out some of her children for a blessing she would not happily extend to all. Thus, the potential to be a God-bearer is present for each of us, insofar as we are willing or able to fully embrace the Divine Presence.

One's inability to embrace the Divine Presence is both voluntary and involuntary. Some consciously choose not to undertake the personal transformation required. Perhaps they doubt their ability to live at such a heightened level, believing that role might require more than they can reasonably or honestly offer. Perhaps because of negative experiences with religion, their image of the divine is not a positive one and they prefer to achieve a higher awareness through humanism or some other secular means where dependence on the divine is not foremost. Such people are to be commended for their integrity and self-awareness. I believe God is joyfully present even in those persons who, because of their life experiences, find it difficult to even imagine God, let alone embrace him.

Others are so wearied and harassed, they are unable to give their full attention to the Divine Presence in their lives. These people are neither to be scorned nor made to feel guilty. Indeed, Jesus felt especially drawn to such people and regularly offered words of hope and comfort to them. Perhaps one day their lives will permit a fuller, more engaged response to the divine. This might well be the case for many of us who find our attentions occupied by the demands of education, employment, marriage, care of our children or parents, physical challenges, or any number of matters that claim our energy and focus.

While some persons make the voluntary decision to not respond fully to the Divine Presence, others are incapable of such a response, burdened with emotional and mental illnesses that compromise their ability to engage the Divine Presence. But if Jesus reflected the divine priorities, we can be confident such people are met with great mercy and compassion.

What we are all called to do, despite our varied abilities to respond, is embrace the Divine Presence *as thoroughly as we are able*. The embrace of some will be tentative, even meager, but it will be the best they can do, and their efforts should be met with appreciation, joy, and thanksgiving, just as the widow with only her mite to give was commended. Others, whose love, soul, strength, and mind are great, will engage God more profoundly. A special few will absorb the divine priorities so fully that they will astound us with their wisdom, grace, and courage.

In a Quaker meeting I once pastored, an elderly woman had committed herself to works of mercy. As I got to know her better, I was astounded at the many ways in which she had blessed hurting people. Though her income was modest, she lived simply so she could give generously. Though her many commitments kept her calendar full, she still found time to be present for those who needed comfort. The longer I knew her, the more I marveled at her charity, given the scarcity of her resources. Because of her humility, she was reluctant to talk with others about her own accomplishments. But one day she let slip the principle that guided her life, when she said to me, "Little is much when God is in it."

I have thought of that many times since, appreciating its truth more and more as the years pass. Little does become much when love is present. Love does magnify our works. Jesus knew this. He knew even the smallest gesture of love could transform the darkest situation and so fully committed himself to divine love that we are still awed by his life. Believe me when I tell you this: we can be like him, and like all the other God-bearers our world has known. It is key to the future of our faith.

Who was Jesus? *One whose awareness of the Divine Presence within him was so keen, and his response to the Divine Presence so full, that he was empowered to live and love so powerfully that those who encountered him were often made whole themselves and more fully equipped to say yes to that same Divine*

Presence that was also in them. We can be like that Jesus. We can be like him when we say yes to the Divine Presence that is also in us, *as thoroughly as we are able.* As we do that, our lives, and the lives of others, will be transformed. God's joy will be in us, and our joy will be full.

5

The Living Spirit

During Lent several years ago, I was invited to speak at a retreat center on the meaning of the Crucifixion. Three of the participants were troubled by a comment I made that called into question the doctrine of blood atonement and asked if they could speak with me to share their concerns. I generally find such encounters interesting, though not always fruitful, but agreed to meet with them.

We met at a local restaurant, and after ordering lunch, I invited them to share their concerns. Only two of the persons spoke, so after a while I engaged the third person, trying to draw her out.

"We decided in advance she wouldn't speak," one of the others said. "She's here to pray for you."

I smiled and thanked her for praying for me, though I had the distinct feeling her prayers were meant to help me see the errors of my way.

"What are you praying for?" I asked.

"We're asking God to give you the Spirit," one of them said. "We all have the Spirit, and it's just wonderful. We want you to have it."

I asked, "What makes you so sure I don't have it?"

This seemed to give them pause, until one smiled and said, "If you had the Spirit, you wouldn't have questioned blood atonement."

"So if I share your theological beliefs, that means I have the Spirit?" I asked.

"Yes," said one, while another, appearing uncomfortable with that conclusion, said, "Not exactly."

While they argued amongst themselves, I finished my lunch and excused myself, leaving them to bicker about the Spirit.

This ambiguity about the Spirit is nothing new. When I was growing up, my family lived across the street from a Pentecostal family who believed one was not truly Christian until the Spirit had caused him or her to speak in tongues. In the Roman Catholic tradition of my childhood, the Spirit's presence and gifts were conferred through the laying on of hands by a priest. The one conviction my Pentecostal neighbors and fellow Catholics shared was their belief that the other did not possess the Holy Spirit.

As a teenager, I was in love with an evangelical girl who told me about the Spirit by explaining the Trinity to me. "Think of it as water, which comes in three forms—ice, water, and steam, but they are all still water. God is one, but has three personalities—the Father, the Son, and the Spirit."*

Anxious to agree with her, I said it made perfect sense, even though it didn't. Unfortunately, my theological agreement didn't increase her affection for me, and after a few dates she left me for someone else. From that experience I learned it seldom profits a man to alter his theology to please another.

The Restriction of the Spirit

Though the Spirit was quite busy in the Hebrew Scriptures—hovering over waters, lending a hand at creation, carrying messages from heaven to earth, helping Israel win the occasional war, and attending to duties in the Temple—it would fall to the church to expand and expound a more developed theology of the Spirit, probably due to the early church's need to explain its ongoing mystical encounters with the Divine Presence after the death of Jesus.

"It is the Spirit he promised to send us," they might have said to one another, echoing the words of John's gospel, when Jesus promised to send his disciples a comforter, advocate, counselor, and helper. This understanding implies the Spirit

---

* I would later discover her view was called *modalism* and deemed heretical by the early church. Modalism asserted that the Father, Son, and Holy Spirit were not distinct personalities, but different expressions of God's self-revelation. Despite her theological error, I was still heartbroken when she jilted me.

was not present with us until after the death of Jesus, a notion pre-Christian communities might have found objectionable. It is a matter of speculation whether Jesus actually promised to send such a presence, or whether those words, woven into the Jesus narrative at least six decades after his death, simply reflected the emerging Christology of the early church.

Whatever the case, it seems clear the early church continued to have mystical experiences they could not otherwise explain except by divine attribution. It is also clear that pattern continued, so that today any experience of the Divine Presence is ascribed to the Spirit, the third person of the Trinity, whose role, and I'm being only slightly facetious, apparently has something to do with one's denominational affiliation. For instance, the same Holy Spirit who routinely inspires a Pentecostal to speak in tongues hardly ever seems to motivate an Episcopalian to do the same. That same Spirit apparently equips only males to serve as clergy in the Roman Catholic tradition, while routinely calling Methodist women to pastoral ministry. It is an odd and capricious Spirit, indeed.

My issue is not with the Spirit, which I have at times experienced myself, but with the church's convoluted and nonsensical attempt in the Trinity, to explain and define the Divine Presence. Why couldn't the church simply have said, "There seems to exist a Divine Presence that is both beyond and within us. It appears some people (Jesus and other God-bearers) are especially sensitive to that Divine Presence and consequently live transformed, even miraculous, lives.

While we don't fully understand this Divine Presence, and perhaps never will, it seems to be calling us toward a life of loving-kindness. This same Divine Presence seems to be no respecter of human boundaries, sharing itself with persons from every tribe, race, and nation." *But again, hoping to contain people's religious experience, the church has qualified it, attempting to corral in cold, precise language a presence that "blows where it will."*

The church's strict parsing of the Spirit's essence and activity has caused the church to persecute those whom the Spirit has inspired differently. Over its long history, the church has spent endless hours bickering over who has the Spirit and who doesn't, where the Spirit is and isn't present, and what the Spirit can and can't do. Ironically, the very Spirit which connects all humanity has been conscripted as a weapon to divide humanity.

The English Quaker, James Naylor, after being convicted of blasphemy in 1656, was whipped through the streets of Bristol, branded with the letter B on his forehead, had his tongue pierced with a hot iron, then was sentenced to two years of hard labor.* This, as you can imagine, had an ill effect on Naylor's health. Two years after leaving prison, he was beaten and robbed. Carried to the home of a fellow Quaker, he died, but not before uttering what has become for many people a compelling and winsome description of the Spirit.

---

* Blasphemy is a term assigned to those whose understanding of the Spirit is different from our own.

There is a spirit which I feel that delights to do no evil, nor to revenge any wrong, but delights to endure all things, in hope to enjoy its own in the end. Its hope is to outlive all wrath and contention, and to weary out all exaltation and cruelty, or whatever is of a nature contrary to itself. It sees to the end of all temptations. As it bears no evil in itself, so it conceives none in thoughts to any other. If it be betrayed, it bears it, for its ground and spring is the mercies and forgiveness of God. Its crown is meekness, its life is everlasting love unfeigned; it takes its kingdom with entreaty and not with contention, and keeps it by lowliness of mind.*

The largesse in Naylor's description resonates with my experience of the Spirit. It seeks only good, working always to reconcile those who are divided. Where cruelty and hatred exist, the Spirit is all the more present, for there its presence is most needed. Having no need to be exalted, nor making any claim for exclusivity, it bears many names—the Spirit of Jesus, the Spirit of Allah, the Spirit of Buddha, the Spirit of the Lord, Elohim, the Great Spirit, the Inward Light, that of God Within Us, the Living Christ, the Holy Spirit, Sophia,

---

* *Quaker Faith and Practice: Britain Yearly Meeting of Friends,* 1995, ch. 19, saying 12. I confess to some skepticism about deathbed statements. If I had just been beaten and neared death, I would not have had the presence of mind to compose such a lyrical sentiment. Whether Naylor said it on his deathbed or wrote it in relative comfort at his kitchen table is incidental. It is a beautiful statement, worthy of sharing.

the Divine Presence, or any other title humanity would assign it, however inadequately that title could capture its essence.

This Spirit was present at creation and abided in Jesus of Nazareth and in every other being who has ever lived and ever will. Because it is life itself, it infuses all life. It need not be asked into our lives, for it is already there. While this Spirit dwells in us, it also transcends us and therefore eludes easy definition. Consequently, it lives beyond the tidy boundaries of creeds and doctrines, no matter how coherent and intelligible those theologies might be.

Obstacles to Knowing the Spirit

There is a man I've known perhaps twenty years. During those two decades, we have found ourselves on opposite sides of several cultural and religious issues. Usually, my disagreement with others hasn't hampered my ability to appreciate and love them. There are many people with whom I disagree whose friendship I nevertheless cherish. But this man has proven to be a thorn in my side, arguing with me bitterly, even hatefully, when we've disagreed. This pattern of division and rancor was one he also established with others, and I would justify my ill will toward him by telling myself others also found him to be objectionable.

Because of my feelings, I avoided him and at times even spoke ill of him. He embodied all the defects I thought wrong with organized religion—arrogance, pridefulness, and rigidity—and none of the virtues—grace, mercy, thoughtfulness, and kindness. I remember even thinking that while I theoretically believed the Spirit was in all people, I found it difficult to believe the Spirit was in him. Now I wonder if the spiritual deficiencies were mine, for my uncharitable thoughts of him had caused me to believe the Spirit had abandoned him when he most needed its redemptive presence. Even as I believed the Spirit could not be present in him, I was living as if the Spirit were not present in me.

At other times in my life, I have taken an excessive pride in my accomplishments, thinking myself superior to others, making me resistant to their possible contributions. I recall one occasion when I wanted the Quaker meeting I pastor to pursue a particular course of action. Though it was a minor matter, I subverted the Quaker process of spiritual discernment, confident I knew better than anyone else the best way forward. Had the Spirit been leading our meeting in another direction, we would never have known, so intent was I on taking us down the path I had chosen. But the Spirit, having no pride, will not insist on its own way. It waits patiently until our pride is quieted.

I remember as a young man meeting a woman who told me the Spirit had called her to work for marriage equality for gays and lesbians. I believed she was dreadfully mistaken and told her plainly the Spirit was not with her. As I reflect on that matter now, I realize the Spirit was indeed with her, that the Spirit had moved well ahead of me on that issue, and that my resistance was an indication of my bias, not the Spirit's. It was, and remains, a strong reminder that we should be hesitant to say to another, "The Spirit is not with you." It might be the case we have not listened closely to the Spirit and have failed to discern the Spirit leading us to a heightened level. That said, let me offer . . .

A Caution About Leadings

I once heard of a man who said he'd been led by the Spirit to sell his home, liquidate his possessions, give his money away, and quit his job in anticipation of the end of the world. When the day approached and passed, he found himself in desperate straits, unable to support himself and his family. I have heard of people who have harmed others, believing the Spirit had called them to rid the world of some perceived threat or evil. While working in a group home for mentally ill adults, I was surprised at how often the residents thought themselves directed by the Spirit to a pattern of life that was ultimately destructive.

When we first resolve to be sensitive to the Divine Presence, we do well to establish the habit of checking our leadings with trusted persons who can be counted upon to speak clearly and honestly with us. Such individuals serve as a check to misinterpreted leadings. On one occasion a woman trained in nursing expressed her sense of being led to move to Africa where she could share her medical expertise. A friend, knowing her tendency to be impulsive, urged her to first volunteer in an inner-city health clinic several days a week. The work proved so satisfying that she eventually left her well-paying job to work full-time at the clinic. The clinic, once beset with leadership problems, blossomed under her oversight. Her nudging had been to help the less fortunate, but her assumption that she would have to go overseas to be useful was inaccurate.

When I sense a leading of the Spirit, I ask myself several questions, including:

Will this bring me joy?

Does the world need this?

Those two questions are inspired by the writer Frederick Buechner, who wrote that our place in the world is to be found "where our deep gladness meets the world's deep need." The intersection of joy and need is often the place to which the Spirit is leading us.

I also ask, "Is this loving? Will this action help my growth and the growth of others?"

I check my motivation, discerning whether I am motivated by fear or anxiety. If so, I wait for further clarity.

I then speak with friends who have evidenced an openness to the Divine Presence in their own lives.

Of course, all of these are no guarantee our discernments will always be correct, but it does protect us from the extremes often associated with religious fervor.

A Closing Word About the Spirit

I grew up in the heart of the Cold War and as a small child quickly adopted the us-vs.-them mentality of my culture. In that worldview, the globe was neatly divided into friends and enemies. Our enemies, the communists, were not like us. They did not believe in the Christian God, did not value human life, and did not cherish freedom or acknowledge human potential. When I was a young adult, the Iron Curtain fell and with it many of the old animosities. In some instances our former enemies became our allies against our new enemies, the terrorists. These terrorists, I've been told, do not believe in the Christian God, do not value human life, and do not cherish freedom. Someday, if history is any guide, our current enemies will no longer be perceived as our foes and a new enemy will rise to take their place.

Unless . . .

Unless we are able to see in one another the Spirit all people everywhere share in common. An evolving Christianity might help us find in one another that which we can love and not fear, that which we can nurture and not kill. It will bridge our differences in language so that when others speak of this Spirit, no matter what name they call it, we will find it to be the very same Spirit in which we all "live and move and have our being." For while it is unlikely we will ever all be joined in one religion, it is not at all unthinkable that we can be joined in one Spirit. If there is a heaven, surely that will be it.

Who Are We?

When our older son was a toddler, it was our custom to go on Saturday picnics. On one occasion, we were visiting a riverside town in southern Indiana. We had eaten lunch at a park, then had walked with our son to a nearby playground where he busied himself on the swings. After awhile, a small boy approached us. He was alone, his parents nowhere to be seen. He was dirty and unkempt. He came within ten or so feet of us but hung back like a stray animal, as if fearing to get too close. My wife and I greeted him warmly and began speaking with him, asking him his name and whether anyone was with him. He ignored us, staring instead at our son, before turning to ask, "How old is your critter?"

Our first response was to chuckle at his choice of words, but we recovered quickly and said, "Our son is two years old. How old are you?"

He told us he was nine, though he appeared much younger because of his slight build and stunted height.

Just as we were about to engage him further, an angry woman, surrounded by a swarm of disheveled children, came into view, cursing at the boy, yelling at him to "get home, if you know what is good for you."

The boy fled, clearly frightened by the woman. The woman turned and stormed away, followed by the children, all of whom appeared similarly cowed by her aggressive behavior.

Over the years, I've encountered many children like that one. Boys and girls who, my wife says, weren't raised but just grew up. Little effort and attention were made to develop their bodies, minds, and spirits. Their childhoods were a feral-like existence. While it is possible to overcome a childhood of indifference, neglect, and abuse, it is difficult and many don't. Instead, they often repeat that same sad pattern, unless they aspire to a more vital, expanded life and commit themselves to growth. That is easier said than done, and many more slog through adulthood with a low opinion of themselves and others, degraded by their childhood experience.

Sometimes such people seek solace in religion. Regrettably, their foray into faith often reinforces the self-image they learned as children. The God they worship bears a stark resemblance to their parents—capricious, ungracious, volatile, and hard. The churches they attend teach them that their primary identity is that of a sinner, of someone who has angered God and has failed to live up to God's high and holy standard. Because that is the self-perception to which they are accustomed, they will believe it, preferring the familiarity of rejection over the unfamiliar feeling of loving acceptance.

Efforts to expand their self-consciousness are often met with skepticism, if not hostility. They fervently believe they do not merit God's love, that were it not for the atoning blood of Jesus, which satisfied God's wrath, they would deserve eternal punishment. This worldview is as pernicious a theology as I have ever encountered, yet it is widespread in Christianity. The combination of a blighted childhood, a broken self-perception, and a degrading theology unite to form a toxic view of humanity. Because this claim can be buttressed by cobbled-together Bible verses, it is seen as authoritative and irrefutable, which makes it all the more difficult to challenge and overcome, especially when it is held by those who view the Bible as God's inerrant Word.

Because of its promise of blessing and threat of punishment, this distorted understanding of faith is not without its benefits. It has been effective in limiting destructive behavior, since some people only seem capable of behaving morally in rigid environments where conduct is strictly codified, rewarded, or punished. Institutions such as the military, prisons, and fundamentalist religion come to mind. But we should never pretend this spiritual rigidity is the ideal goal of faith, and we should always encourage one another to move beyond this stunted stage as our God-awareness and self-awareness grow.

Unfortunately, some people never advance but remain in that constricted stage their entire lives. Indeed, some never make even that much moral and spiritual progress, which is one reason prisons are full. But as our understanding of God and Jesus expand, our self-understanding will also change. We will no longer believe we are moral pygmies incapable of decency and deserving of God's wrath. Instead, we will see ourselves as capable of great good, filled with transcendent beauty and rich promise, indeed filled with the Divine Presence with the same potential of Jesus and other God-bearers. When we perceive that, our behavior will follow our perception, and believing ourselves to have dignity, worth, and value, we will treat ourselves and others with dignity, worth, and value.

Sinner or Saint?

Several years ago, I was speaking at a church where the worship service was concluded with an infant baptism. During the ceremony the pastor spoke about the child's being born in sin. As he spoke, I caught a glimpse of the child. His right hand was curled around one of his mother's fingers, a beatific smile illuminated his face. Were I asked to paint a picture of an angel, I would have painted that infant. I wanted to stop the ceremony and ask the minister how that tiny, beautiful infant could possibly have sinned. Think of it! The church's first liturgical words to that child were to condemn him.

To be sure, if I had paused the ceremony and asked those gathered if they actually believed the child were evil, many present that day would likely have said, "Of course not." Yet those words and images persist in the Christian tradition, having taken root after centuries of repetition, so pounded into our collective consciousness that they are uncritically accepted as gospel truth by many in the church and culture. *Indeed, in most Christian traditions, the path to God and salvation begins with the explicit, and often public, acknowledgement that our nature is primarily wicked.* How odd that an enterprise whose goal is the enrichment of humanity should have as its core belief the conviction that we are depraved.

I was once in a grocery store, standing in line behind a man and woman. The woman, realizing she had forgotten an item, sent her husband in search of it. He returned a few moments later with the wrong item, causing her to snap at him, saying, "Can't you do anything right?" In defense of himself, the man replied, "Hey, cut me some slack. I'm only human."

Of course, I had heard that statement many times before and in fact had used it in my own defense. But for some reason, in that instance, the phrase, "I'm only human," intrigued me and I thought about it at length. Behind that statement lay the inference that to be human is to fail. More telling, the use of the word "only" suggested that we humans are, by our very nature, deficient. We are "only" human, as if being human were a fault to overcome, not a condition to be celebrated. I suspect that in all the animal kingdom we are the only species to believe its inherent nature is flawed.

The danger of this is our tendency to live up to, or down to, our billing. I once attended a trial with a man and woman whose son had been arrested for a felony. The parents and I sat together for several hours, watching other cases tried while waiting for their son's turn. When his case was announced and he arose to face the judge, his mother leaned over to me and said, "I knew this was going to happen back when he was a little boy. I told him then he'd end up in jail someday."

Instinctively, without thought, I replied, "If you tell a child he'll end up in jail, you shouldn't be surprised when he does."

"I'm not surprised," she said, the irony of the situation escaping her.

Over the next several years, I witnessed the pessimistic manner with which this woman treated her children, belittling their accomplishments and predicting all manner of failure for them. Her persistently gloomy attitude acted as a toxin in the family, creating hard feelings and ill will. Predictably, her children struggled, burdened by crippling self-images. Oddly, their mother seemed delighted by their failures, as if it confirmed her assessment of them.

Whenever I hear a Christian leader trumpeting human sin, I remember that woman and the malignant effect of her words and wonder how the church's poor judgment of us can't help but achieve a similar result of failure and self-hate.

When our primary identity is that of sinner, we will invariably descend to that low calling.

I want to contrast that negative view with a touching message I once heard delivered in a Canadian church. I had been invited to speak at a progressive Lutheran church in Toronto. The church had been the subject of some controversy when it became widely known that the minister, a young woman, was a lesbian. To her credit, she had been open and honest about her sexual orientation, and to the congregation's credit, they recognized and affirmed her obvious gifts for ministry and employed her as their pastor.

It was All Saints Sunday, and during the course of worship, the pastor invited the children present to come forward. She explained to the children that it was the day the church celebrated all the saints who had populated the church over the centuries. The minister asked the children if they knew what a saint was. One little girl said, "A saint is someone who is especially special, who God loves especially lot." While her answer wasn't grammatically correct, we knew what she meant and found it very touching. Other children pointed out that saints were old, mostly men with beards, that they were kind and smiled and helped people.

Then the minister told the children that coincidentally, there just happened to be a saint with them in church that very morning, an honest-to-goodness saint she wanted them to meet. The reaction of the people was very interesting. The children stood up and began looking out in the congregation trying to find the saint, and the adults began looking at one another, wondering if the saint were seated next to them.

When everyone had finished looking around, the minister reached into the folds of her robe and pulled out a small mirror, which she held before each child, saying. "There's the saint." The children and adults laughed and smiled, delighted. It was one of the holiest moments I'd ever experienced in a church, one that stayed with me, and one I'm confident will stay with those children and their parents.

These should be the church's first words to children, not the accusation that they are sinners, but the affirmation that they are saints. For we live up or down to our billing. That has always been the case and always will be. Our sense of self is the product of other's perceptions of us, our awareness of how others see us. Our self-perception is formed early in life, is perhaps one of the earliest traits formed, and the most difficult to "rewrite" once it takes root. Indeed, it can be so entrenched that it becomes determinative, shaping our lives for good or ill in ways we never imagined.

While a teenager, I read a magazine article about the composers and brothers George and Ira Gershwin. In the article, as best I can remember, George said of his brother, "Ira always believes everyone will treat him honestly, so they do. I always believe everyone will cheat me, so they try." Because we so often embody the expectations others have of us, we ought to pay far greater attention to what the church says about humanity. *The church's negative assessment of humanity, rather than assisting in our redemption, has caused untold psychological and spiritual damage to millions of people.* It is long past time a new story of human worth, dignity, and potential was told.

The Problem with Pessimism

More than anything else, negative religion rejects an optimistic view of humanity, for negative religion hinges on the belief that humanity is fallen and evil, in need of a savior who intervenes in history to save us from our sin. How odd that the foundation of any religion should be the utter rejection of our innate capacity for good.

I remember touching on this subject during a speech I was giving at a respected Midwestern university. Members of the local chapter of the Campus Crusade for Christ* had turned out in high numbers, and after my remarks they hurried to the audience microphones in an effort to dominate the question-and-answer session that followed. Though the purpose of the exercise was to engage in conversation, it quickly became a "revival" atmosphere, as student after student stood and rejected a positive view of humanity, disparaging our capacity for good, urging those present to turn from sin and accept Jesus so they could be saved.

---

* The Campus Crusade for Christ is a college-centered, fundamentalist ministry whose purpose is to influence future leaders. Among its stated beliefs is this statement: Man's nature is corrupted, and he is thus totally unable to please God.

It soon became clear our hoped-for conversation couldn't happen so long as they dominated the microphones, so I asked them to nominate one of their peers to ask three questions, which I would answer, then give others of a different theological bent an opportunity to ask questions or make observations. During the next hour, I watched the give-and-take between two groups of persons whose views were radically different. Curiously, I observed that many who argued for a positive view of humanity were elderly, had lived through the Great Depression and World War II, or had witnessed many tragic and evil situations yet still bore eloquent testimony to the human capacity for loving kindness. The younger people had been the recipient of parental love and concern, had never known want or war, enjoyed the blessings of advanced health care, and were attending one of the finest universities in the nation but were utterly convinced the humans who had sacrificed and worked to make all those blessings possible were intrinsically depraved.

Coincidentally, I would later meet the parents of some of these students, many of whom expressed dismay at their children's conversion to fundamentalist faith.

"It was as if they targeted my son," one mother told me. "It was his first time away from friends and family. He was lonely, then along came a group of instant friends who convinced him the religion he'd grown up with was wrong."

I can readily understand how those who are needy, who have suffered evil, loss, and want, can become pessimistic and jaded. What I do not understand is how those who have known material and familial blessing could be so cynical as those young students. *The only possible explanation for such irrational fanaticism is the unthinking fervor integral to negative religion.* Because their worldview is rooted in religion, because it strikes at the heart of what it means to be human, their passionate belief in humanity's evil is considered a virtue, not a vice. That is, the more they believe in their own depravity, and the depravity of others, the more they will be applauded for their devotion to God. *This is the inverted nature of negative religion—it applauds what should be soundly rejected and rejects what should be roundly affirmed.*

Worse than that, pessimistic religion, by its very nature, is ill-prepared to help humanity deal with our most intractable problems. Because negative religion views humans as intrinsically evil and ultimately incapable of good, it offers little hope for our future, save for the intervening action of an all-powerful God, who, if you will forgive me for saying so, seems curiously unable or unwilling to insert herself in time and space to rectify our problems. *In short, pessimistic religion is quick to point out our failures, hastens to remind us of our inability to solve them on our own, then demands we place our faith and trust in a deity of its own invention, whose track record for alleviating human suffering is spotty.* Why the moral, logical, and spiritual bankruptcy of this dismal worldview is not readily discerned and acknowledged by otherwise intelligent persons is beyond me.

The rejection of human goodness, resourcefulness, and power, coupled with a belief in a "celestial rescuer," hinders our ability to face our problems squarely, assess our challenges realistically, and overcome them creatively and decisively. In that sense, negative religion is the enemy of human progress. Those who hold such views should be treated graciously, but their religion should be recognized for what it is—an outdated, implausible, superstitious faith, whose contributions to the modern world are negligible, whose harm is substantial. Whenever it dominates a culture, people are diminished, progress is scorned, and prejudice is sanctified, all in the name of God, whose goodness comes always at the expense of human virtue. By this I mean that for God to be a rescuer, humans have to be fallen and corrupt, in need of a savior. Thus, in negative religion, it is logically impossible for both God and humanity to be simultaneously virtuous. Those who believe that religion should uplift our spirits and expand our minds should soundly reject any religion that asserts such a bleak and twisted view of human nature.

The Value of Optimism

When I first became a Quaker, I observed the children in the meeting calling the adults by their first names. I was a teenager, still a child myself, and accustomed to addressing my elders as Mr. or Mrs. But when I did that among the Quakers, I was told those honorifics weren't necessary, that I should feel free to call the adults by their first names. Though I didn't know it at the time, this was a long-standing custom in the Society of Friends, borne out of our testimony on equality, which held that all persons, regardless of their social status, were to be equally esteemed.

One Saturday morning, my father and I were at the post office and encountered an older Friend in the lobby. He greeted me, and I returned his salutation, greeting him by his first name, Marvin. We exchanged pleasantries, then Marvin left. My father turned to me and said, "You should call him Mr. Rutledge."

"That's not the Quaker way," I said. "We call one another by our first names."

This perplexed my father, who wanted to support my religious choice but was mystified why any religion would encourage what he thought to be a breach of good manners.

Though I was too young to articulate the value of this custom, and perhaps even too young to fully appreciate it, I do recall feeling a heightened sense of responsibility by this gesture of equality, wanting to be worthy of the esteem shown to me by these Friends. My father's fear that I might have less respect for my elders never materialized. Just the opposite was true—because I valued the manner in which I was treated, I felt doubly motivated to return their courtesy, striving to warrant the honor in which I was held.

After graduation from high school, I moved out of my parents' home into my own apartment. Like some young males,

I responded to my newfound freedom irresponsibly. Even though I behaved recklessly, I still made the Quaker meeting a part of my life. Those gentle people were probably mystified by my behavior, and perhaps even deeply concerned, but I was never admonished or made to feel guilty. Instead, they graciously welcomed me into their homes, where I witnessed firsthand the depth and richness of their lives. I eventually realized the path I was on could not possibly end well. Inspired by their example, I resolved to aim higher, devote myself to wiser endeavors, lest in my irresponsibility I make a mistake that could damage my life and the lives of others. A year or so later, I met my future wife and was by that time sufficiently mature and ready for an enduring relationship. Had I met her earlier, I am almost certain she would have refused to date me.

All of this is to say that the Quakers' confidence in my innate decency was eventually life changing. Had I been scorned, had their treatment of me been negative, I might well have spiraled downward. Instead, I was motivated to prove myself worthy of their respect, so gradually did.

This optimism in human potential was illustrated well in the Bible, in the nineteenth chapter of Luke's gospel, when Jesus treated Zacchaeus, a notorious cheat, with dignity and respect. Though Zacchaeus had done little to merit anyone's optimism, Jesus assumed his potential for good, and in that act of grace he inspired Zacchaeus to turn a new way. The response of those who knew Zacchaeus was all too typical. They were scandalized and angered by Jesus's fellowship with a known sinner.

Perhaps the greatest obstacle to human transformation isn't our inability to change but the unwillingness of others to believe our transformation is possible.

It seems odd that religious people, who are usually insistent on the need for human transformation, are sometimes the most resistant to its likelihood.

Several years ago, an acquaintance of mine was struggling with alcoholism. I had spoken with him several times, to no avail, and had given up hope he could turn his life around. When he was at a particularly low point and had lost his job as a result of his alcohol abuse, I was talking with some mutual friends who had resolved to help the man. They were confident of his potential, were willing to invest time and effort to help him, and had come to me asking for my participation. I was reluctant to do any more than I had done, believing the man's chances for a changed life were slim. None of the others seeking my assistance were particularly religious. So while I was the only one explicitly committed to a life of faith, I was also the most pessimistic, the one most resistant to the possibility of transformation.

I've since pondered the correlation between religion and optimism, whether there is something intrinsic to religion that makes people of faith skeptical of human goodness and potential. Is it because we have grown so accustomed to the incessant drumbeat of sin that we find it difficult to believe in our capacity for decency?

Thankfully, my friends did not let my cynicism keep them from assisting this man. They urged him to seek treatment and provided transportation to AA meetings. He in turn committed himself to sobriety, has abstained from alcohol ever since, and is happily employed and restored to wholeness. As for me, I am profoundly grateful that my friends hadn't given up on the possibility of new life.

Optimism as a Spiritual Virtue

I was once speaking on the topic of spiritual optimism, and during the question-and-answer session a man stood and accused me of being naive, offering a litany of evidence confirming humankind's depravity. While it isn't my custom to argue back, I invited the man to consider the difference between naïveté and optimism. Naïveté has to do with being uninformed and inexperienced. In matters of the human condition and faith, I am neither. I'm fifty years old, have pastored over half my life, and am well acquainted with human sin and evil, my own and others'.

I am not naive. But I am optimistic. In my life, I have witnessed countless acts of grace, have seen many lives transformed by the power of good, and have observed evil and hatred overcome by mercy and decency. I can count on one hand the persons I have met whom I believed were truly evil, whose wickedness could not be explained by maltreatment. But I could fill entire books with the names of persons whose charity and compassion were deeply inspiring.

Spiritual optimism is a recurring theme in the lives of the deeply spiritual persons I've met. They not only believe the best things about God, but they are convinced of humanity's capacity for good. While not blind to human failings, they remain positive about our progress and potential.

Many years ago, I was visiting an older gentleman. A retired pastor, he had served several congregations well and faithfully, helping them become rich, life-giving communities. He was over ninety years old but still possessed a bright, quick mind. I would drop in on him occasionally to engage in theological discussions.

"So," I asked him, when we were settled into our chairs in front of his fireplace, "were we born sinners?"

We wasted little time on niceties and usually got right to the point.

He grew quiet, thinking, then said (and I'm paraphrasing because our conversation was held so long ago), "I believe in evolution. I believe each species adapts to its environment in order to better thrive. We aren't born sinners. We are born incomplete and immature. The goal of each generation is to advance our human growth and evolution. It's a long process, but an unstoppable one."

He went on to discuss the progress in the span of his lifetime, citing the growth in knowledge, the awareness of our global connectedness, and the efforts to create governments that benefited many, not the few.

"It is a slow process," he said again, "but if you compare where we are now with where we were several hundred years ago, you can't help but be an optimist."

I challenged him, citing wars and greed and human suffering. He readily acknowledged those things but held to his conviction that humanity was moving in a positive direction, that our concern about war, want, and selfishness was itself proof of an evolving morality.

During the course of our conversation, my worldview shifted. Because I was raised hearing the church's steady drumbeat of human fault and failure, I had assumed our corrupted nature was an undeniable fact, the necessary starting point for Christian theology. I had believed that was an indisputable element of the gospel and could not be questioned. But listening to this wise man was a revelation to me. *We were not born into sin. We were born incomplete and immature, slowly and inevitably working our way toward a more evolved state.* Gone was the judgment, the urgent need that our first words to God be apologetic, the abject sense of failure. All of that was replaced with a warm gratitude for the spiritual and moral progress of our ancestors and a healthy desire to creatively further our ethical evolution. This proved to be a spiritual breakthrough for me. My faith, which had often felt like an onerous, relentless burden, became infused with peace and joy and lightness.

Around the same time, I happened upon a quote from the Quaker writer and teacher Elton Trueblood, who said, "Pastors ought always be working themselves out of a job." I presumed that Trueblood meant that pastors ought to be equipping others for lives of compassion and grace, and to be doing that so ably that their own presence would no longer be required in a specific community.

When my children were young, it occurred to me that the work of a parent is much the same—to equip our children for life so that our presence, while still appreciated, will one day not be crucial for their own functioning.

Perhaps this is also true for God. When we perceive ourselves as sinners in need of salvation, then God will always be needed to rescue us. We will always require a god to save us from our sin. *But if God is the universal and inward impulse that inspires us to seek the best for others and the growth of the beloved, then when we have learned to do that for ourselves, when we have evolved to that fuller awareness, then God might well have worked herself out of a job.* Perhaps when we are fully evolved, we will no longer require a deity's assistance, for we will embody the attributes and goals toward which God has called us. We will be, in every sense of the word, grown-up.

Is it too scandalous to believe that a spirit of optimism might well lead us not only toward a fuller humanity, but toward a fuller divinity, when we need look no further than within to see the face of God?

Suffering and Evil

7

When we were first married, my wife and I rented an apartment from a mortician, who had purchased the century-old house next to his funeral home, converting it into three pleasant apartments. We became friends with the funeral director, a thoughtful, caring man named Norwood. I had just begun attending college to prepare for my vocation as a pastor but hadn't yet had any experience as a minister.

One evening Norwood phoned our home, explaining that a young couple had lost their child in an accidental death. The couple were without a church and didn't have a family minister, so Norwood asked if I might be willing to conduct the funeral service for their little boy. Though I had never performed a funeral, I wanted to help the mother and father and agreed to do it.

I chose as my scripture reading the 121st Psalm from the Hebrew Scriptures.

I lift up my eyes to the hills. From whence does my help come? My help comes from the LORD, *who made heaven and earth. He will not let your foot be moved, he who keeps you will not slumber. Behold, he who keeps Israel will neither slumber nor sleep. The* LORD *is your keeper; the* LORD *is your shade on your right hand. The sun shall not smite you by day, nor the moon by night. The* LORD *will keep you from all evil; he will keep your life. The* LORD *will keep your going out and your coming in from this time forth and for evermore.*

I had committed much of the psalm to memory, so I was watching the mother and father as I spoke the words, "The LORD will keep you from all evil." The instant I spoke those words, I regretted them. The reaction of the mother and father was palpable. They seemed incredulous. I suspect only their good manners kept them from blurting out, "What do you mean the Lord will keep us from evil? What do you call this?"

After the scripture reading, I tried to soften its impact, observing that sometimes it doesn't feel as if God protects us from evil, but it was a failed effort to defuse this spiritual bomb dropped into their laps, and I had the distinct feeling they wanted nothing to do with a god whose promises rang so hollow.

In the years since, I have seen this scenario played out time and again, the thoughtless recitation of words meant to comfort that have had the opposite effect, leaving grieving persons doubly devastated by the realization that God not only failed to protect them, but might even have been complicit in their suffering. As a consequence, they wanted nothing to do with God or those who believe in God.

Because I have found church such a rich experience, I'm intrigued by persons who want nothing to do with it and often ask them why they choose not to be involved in a Christian community. Once they sense I'm not trying to convert them or make them feel guilty, they often speak candidly about the barriers that keep them from participation. They frequently cite the problem of suffering and evil, wondering how an all-powerful, all-loving, and all-knowing God can let suffering and evil go unchallenged. Many of them grew up in the church where they heard about a God who intervened to save believers from evil, pain, and suffering. This tenet held up until they experienced evil, pain, or suffering; then they began to wonder what other untruths the church had taught them. Or perhaps they continued to believe God had the power to spare them from hardship but didn't, causing them to resent God. Eventually, either their skepticism or bitterness prompted them to leave the church.

Those of us in the church sometimes suppose that people who don't attend are spiritually apathetic or sinful. But perhaps their integrity keeps them away. Perhaps the god the church describes is one they no longer believe in, and rather than pretending to a piety they don't possess, they elect to stay away. This is all the more reason for the church to speak honestly about God's role in suffering and evil—that God lacks the power to spare us from hardship despite our fervent wishes that it were otherwise. *Our continued insistence that God can save us from evil is not only dishonest, it discourages ethical persons from participating in a spiritual venture they believe isn't altogether truthful.* Simply put, what the church has long claimed about God isn't true. Praying in faith doesn't always lead to physical healing. Praying for divine protection doesn't ward off evil or hardship. No evidence exists that suggests those who believe in God suffer calamity any less than those who don't. Nor, despite the claim of almost every religion, does participation in a particular faith assure a supernatural protection unavailable to those outside it.

The Universal Wish

Despite the outward differences in the world's religions, as one begins to study them, one notices several commonalities. Religions often begin as an effort to either explain suffering or protect against it. The first book of the Bible is barely underway before the writer is proposing a cause for suffering and death. "And the LORD God commanded the man, saying, 'You may freely eat of every tree of the garden; but of the tree of the knowledge of good and evil you shall not eat, *for in the day that you eat of it you shall die*'" (Gen. 2:16–17).

Myths and stories that try to make sense of the mystery of suffering are found in nearly every religion. We don't seem

nearly as mystified by the presence of happiness and blessing, which makes me wonder whether suffering and evil
might be the prerequisites for religion. That is, if humankind
didn't know hardship, would we be nearly as motivated to
create systems of faith? In my case, the death of my closest
childhood friend drove me back to the church, in an effort to
make sense of his senseless and sudden passing. Had he lived,
had I never encountered such deep loss at such an impression-
able age, I'm not certain I would have engaged religion to the
extent I have.

This effort to seek meaning in the midst of sorrow might
be a universal impulse, and our hope of avoiding pain by
seeking God's favor, a universal wish. So deep is our wish for
heavenly intercession, we often resist assertions that call such
intervention into question. Shortly before this book was written, I was invited to speak at a church in the south. Located
in a university town, the church was populated by persons
affiliated with the school, infusing it with an intellectual and
spiritual energy I found appealing. They were quite receptive
to my belief in universalism and enthusiastic when I spoke
about the progressive beliefs we held in common. Afterward,
during the question-and-answer period, a woman asked me
if I believed in angels. I thought for a moment, then said
that while I'd met people who were angelic, I had never encountered the kind of angel described in the Bible. I further
said I'd seen no convincing evidence that heavenly beings
regularly intervened to protect us from harm. I concluded
by saying it would be nice if it were true, but if it were, I'd

want to know why angels saved some people and not others. As is my custom, whenever I sense I might be stepping on someone's spiritual toes, I said this pastorally, not in a flippant manner.

Nevertheless, the woman was insistent that she, and everyone else, had a personal angel whose job it was to protect each one from suffering and evil. A few weeks before, I had conducted the funeral for a young woman killed by an impaired driver. I briefly shared that story with the audience and asked whether the parents of the young woman would agree that each of us had been assigned a protective angel. But the questioner remained adamant, saying she and countless others had been saved by their guardian angels. "Why don't you believe that?" she asked me.

I replied, "I don't doubt that you believe in angels. What I doubt is whether angels actually exist, and if they exist, whether they protect everyone. Too many people suffer too greatly for me to believe in the universal, protective entity you're describing."

Though my statement seemed entirely reasonable, at least to me, she and several others in that church shook their heads in objection. At the conclusion of my talk, they gathered around, urging me to believe in angels. I reminded them that a key component of healthy spirituality is speaking honestly from one's own experience. In my experience, I had seen no evidence of angels, or any other divine agent acting to save us from suffering, and therefore I was unwilling to assert their existence. I hastily added that they were quite free to believe in such beings. But even that was clearly not enough, and they left the church disappointed and upset.

I've since concluded that for beliefs in divine or heavenly power to be sustained, for them to be comforting, it is imperative that no one call them into question. For too many people, the prospect that God or angels or Jesus or Allah or Elohim doesn't or can't intervene to protect us from suffering and evil is ground shifting and consequently terrifying. *But any religious principle based on fear, unreality, and avoidance is doomed to fail, ultimately unable to avert the very misfortune it was designed to prevent.* Nevertheless, these spiritual tremors should not be feared, for most theological and spiritual shifts occur not after great blessings, which we usually assume to be our due, but after great sorrows, which have a way of overturning our theological assumptions. These shifts often result in deeper joy, honesty, and awareness, despite the initial pain that often accompanies them.

Consider this great irony: the very pain and suffering we'd hoped to avoid by becoming religious is often the very means by which our spiritual wholeness, integrity, and happiness become possible. It is no accident that some of the happiest people I've met have been those who regularly insert themselves into situations of great suffering in order to bring healing. Some of the unhappiest persons have been those who made pleasure their goal, strenuously avoiding all difficulty, pain, and brokenness, but were at the end of the day the most dispirited and empty.

The Necessity of Facing Suffering Honestly

A number of years ago my telephone rang late in the evening. It was a man I had met a few months before, phoning to tell me he was alone in a hotel room, had gotten drunk that day, and needed help. Having tried to help alcoholics in the past without much success, and having spent the past few days away from my family, I was reluctant to peel myself away to be with him, suspecting he wasn't serious about changing his life. Nevertheless, I went and sat with him in his hotel room while he talked about how difficult life was and how the only peace he felt was when he drank.

"But when I sober up," he said, "my life is even worse. I've lost my job, my house, and my family. Not even my parents will help me."

I spoke frankly, even brusquely, with him, hoping to jar him into treatment. After an hour or so, he agreed to let me drive him to an alcoholic rehabilitation center, where he checked in for treatment. He lasted two days before checking himself out and getting drunk. He phoned me again, asking my help. I declined to meet with him, believing it would do little good to help him until he was motivated to be well.

For the next year or so, he would phone me at irregular intervals, asking for money. Though it was difficult, I told him no, that I would only assist him when I saw a sincere commitment on his part to stop drinking.

One morning he called, sounding clearheaded. "The Lord

saved me," he told me. "He took away my urge to drink."

He told me he'd met a Christian who worked with alcoholics and addicts, that the Christian had taken him to his church, where the pastor had laid hands on him, curing him of his addiction. He'd been baptized, joined the church, and was now helping other alcoholics. I expressed my happiness for him and wished him every success.

Because the church was a large one and not far from my home, I was familiar with its pastor and history. I knew this pastor had a tendency to overpromise the blessings of God, guaranteeing his congregants financial prosperity, physical healing, protection from "Satan's attacks," and liberation from drug and alcohol addictions. In short, he promised them their lives could be free of every pain, that their world could be one in which tears never flowed, if and when they "accepted Jesus as their Savior and were anointed with the Holy Ghost."

This had apparently resonated with the man, who'd spent much of his adult life trying to avoid pain and suffering and had now been virtually guaranteed God would smooth his rough path. Though I had my doubts about his recovery, I had been wrong many times in the past, witnessing transformations I never believed possible. I hoped for his sake he was truly healed and thought of him often, wanting the best for him.

I didn't hear from him or about him for several months, then learned from a mutual acquaintance that he'd begun drinking again after the death of his mother. His mother had been hospitalized, he had gone to visit her and had kneeled at her bedside praying for her recovery, confident she would be healed. But she was too old, her body too worn and sick, and she passed away, plunging him into a crisis of faith, to which he responded by drinking.

Shortly after that, my family and I moved to another town and I lost track of the man. But I think of him from time to time and wonder whether he is well. More specifically, I wonder whether he has learned to face his suffering directly, without the anesthetizing effect of alcohol. Or, for that matter, the anesthetizing effect of religion.

What do I mean by facing suffering directly?

I mean that many of us are preoccupied with our own happiness, spending vast amounts of time, money, and energy in an effort to avoid suffering of any sort. But because suffering is inevitable, because it comes to each of us eventually, our unrelenting efforts to escape its grasp invariably fail. At best, we can postpone suffering, but even then it has a way of multiplying like a malignancy until it is acknowledged and addressed. Nevertheless, the belief we can somehow evade suffering is pervasive and persistent.

Not long ago, I was standing in the checkout line at our local grocery store, waiting my turn, scanning the covers of the magazines for sale and noticing a recurring theme. Almost every magazine featured a story or column advising its readers how to be happy. Though I didn't have time to read the articles, I suspected their advice was generic, given the inherent limitations of glamour magazines. Yet the lure and appeal of happiness is strong, evidenced by the popularity of this theme in books, magazines, and TV talk shows. To desire happiness is natural. None of us, after all, wishes to suffer. But our *desire for happiness* can grow into an *expectation of happiness* when we begin to think happiness is our due, something owed to us by God or the universe or others.

An acquaintance of mine, previously indifferent to spiritual matters, joined a church and became fervently religious. Shortly after his conversion, he was laid off and had difficulty finding new employment. As the weeks passed, he grew angry and said to me, "God isn't keeping up his side of the deal. He said if I put him first, he'd take care of me."

Of course, God had said no such thing. A television preacher, urging people to send money to his "ministry," had assured those giving that they would have every need met. The man, theologically unsophisticated, believed the preacher, sent him money, and joined a local church, anticipating a windfall, only to experience continued need.

The church is not the only arena in which happiness and blessing are promised. Billions of dollars are spent each year on drugs, both legal and illegal, psychiatrists and psychologists, therapists, medical doctors, alcohol, stimulants, food, vacations, real estate, cars, jewelry, electronics, books, clothing, and a myriad of practices and possessions, in hopes they will make us happier or, at the very least, postpone our suffering. This makes our suffering all the more difficult to bear, since we had invested time, money, and energy avoiding it, only to discover our efforts were ultimately fruitless.

When the avenue to happiness is a religious one, we also experience, in addition to our deep frustration, a sense of abandonment, believing God has forsaken us, leaving us to our own defenses. The people we love still die, we still experience deprivation or illness or depression. What is often lost in this experience is the recognition that while God might be unable to prevent our suffering, our fellow humans can be an invaluable help in negotiating the terrain of loss and sorrow.

My wife, Joan, and I delayed having children until I had finished my formal education. When we learned Joan was pregnant, we joyfully began preparing our home and lives for our first child. Unfortunately, in the middle of the pregnancy, our baby died in a spontaneous abortion, or miscarriage. We had waited years to have children and were numbed by our loss. At the time, we were involved in a Quaker community in Indianapolis. As word of our loss began to circulate through our meeting, people arrived at our home with food, flowers, and words of consolation. Though some of the comments were less than edifying—*This is part of God's plan. God needed an angel. One day you'll understand why*—we deeply appreciated the kindness behind the words and knew they were intended to help, not hurt.

But the compassion was not limited to those in our church. Family, neighbors, and coworkers treated us with deep grace.

They took our loss seriously, but appropriately, helping us gradually place our loss in its proper context. In short, they did for us what loving people should do—they tenderly assisted us when we were crippled, then empowered us to stand and walk again.

God's Hope for a Connected and Capable Humanity

While there is little evidence that God intercedes dramatically and dependably to eliminate human suffering, there is regular confirmation that our fellow beings can and do bear us up when we cannot stand alone. Perhaps if God does "intervene" in our lives, this is how she does it, by inspiring persons sensitive to divine promptings to make themselves available to those who hurt. And perhaps God also inspires us toward greater self-sufficiency in the interests of our personal growth and maturity. Might this have been God's method all along, so that we would look less to her and more to ourselves and one another?

When I was growing up, there were two families in our small town who lived side by side. There were three children in each family, roughly the same ages. As the children grew to adulthood, some interesting differences began to emerge. In one family, the children never quite found their footing, returning to their parents' home to live after failed marriages or economic setbacks, which seemed to occur with startling regularity. Even as the adult children neared their fifties, they continued to rely upon their parents. For their part, the parents did nothing to discourage their children's dependence, giving them whatever they wished, never insisting their children become self-sufficient. Not only did the adult children continue to remain financially dependent on their parents, they were also emotionally reliant, constantly seeking their parents' guidance on matters large and small, even involving their parents in their marriages, which, as you can imagine, created much ill will with their spouses.

In the other family, the adult children launched from the nest after high school. Two of the children attended college, one did not, but all three quickly moved toward emotional and financial independence. The adult children married well and happily. The children, as is true with all people, made some poor choices but did not expect their parents to bail them out, nor did their parents offer. Consequently, the children learned from their mistakes, enjoyed the satisfaction of overcoming adversity through their own ingenuity and initiative, and were better equipped to handle the challenges of life.

Eventually, the parents died. The children in the first family fell apart, unable to care for themselves, either emotionally or physically. The children in the second family mourned the loss of their parents, but moved forward with confidence and competence, continuing to succeed in their relationships and vocations.

Now let us think for a moment about those persons who lurch from one crisis to another, who find themselves in a constant state of suffering and difficulty, and speak often of their complete reliance upon God.* Let us ask ourselves, "Is God honored by their constant and utter dependence? Or is God most pleased when we live with competence and maturity, not requiring God's attention and assistance in every matter?"

As the parent of two sons, I am pleased when my children are self-sufficient, even though their personal resourcefulness means I am needed less and less for their everyday functioning. Indeed, short of a serious physical or mental impairment, I would be disheartened were my sons to continually depend upon me for their well-being. I would feel as if I had failed in my duty to properly prepare them for life. I don't say this with a sense of smugness, for my children are still young, not yet launched from our nest, so their future abilities are undetermined. But my hope for them is a maturity that enables them to negotiate life successfully without my constant and continued intervention.

This isn't to say they won't need other persons in their lives. Humanity is, after all, connected. We never lose our need for companionship, empathy, wisdom, understanding,

---

* I do not mean to include in this group those persons who are physically or mentally incapable of caring for themselves. Nor do I mean to include those who have been born into such dreadful circumstances that no escape or improvement is possible.

and all the other gifts people offer us. But it is to say their utter physical and emotional dependence on a parental figure will diminish as they mature. Indeed, I would hope that as their personal need for a sustainer lessens, they will serve that role for others, perhaps their own small children, who in turn would also one day move from dependence to self-sufficiency.

The Cycle of Life

Near the end of her life, my mother-in-law, Ruby, moved from the family farm in southern Indiana to live in a retirement home in our small town. Because she had lived in her farmhouse for nearly sixty-seven years, she was reluctant to move, but after a series of health setbacks, including memory loss and a few near accidents, her adult children insisted she move to safer housing for her well-being. Prior to Ruby's rather sudden decline, she had been in full control of her faculties, able to make appropriate decisions, tend to her personal needs, and fully engage others.

Because her new living quarters were so close to our home, it was possible for my wife and me to visit her each day. Ruby looked forward to our visits, greeting us warmly when we arrived and always ending our visits by inviting us to return the next day to see her. As the months passed, her motor skills declined, so our participation in her life increased, helping with laundry, bathing, and eventually walking. Because Ruby was modest, my wife assisted in the more private aspects of her care, extending to her mother the type of support she herself had received as a small child. Life had come full circle for Ruby and my wife, as it eventually does for each of us.

When Ruby passed away, I began thinking more deeply about the rhythm of life and its correlation with human spirituality, specifically about our relationship with God, who is often understood and described as a parental figure. I wondered if our relationship with God mirrored our relationship with our parents, moving through the cycles of dependence, mutual friendship, and reciprocal assistance?

The stage of dependence happens when we are young or spiritually immature, or in crisis or need. During that time, we look to God primarily as a parental figure who can be trusted to provide for us and protect us from life's hazards. We believe God does for us that which we believe we are unable to do for ourselves. We might say, "If God hadn't been there for me, I wouldn't have made it." Our orientation to God in this dependence stage is twofold—we ask God to intervene in our lives in very specific ways, and in turn we praise and thank God for his presence, power, and compassion. In this stage, God has the ability to end our suffering and sometimes does (at least we attribute our rescue to God), though sometimes God permits us to experience suffering in order that we might learn a great truth or achieve a higher, as yet unknown, blessing. God does for us what our parents did for us when we were children—God protects us, God provides for us, and though God could protect us from all harm, he is wiser than us and sometimes lets us experience pain so we might learn a lesson or later enjoy a greater good. This interpretation of God and suffering is one I have often heard expressed in more traditional Christian circles. A majority of the church's hymns reflect this worldview, and it enjoys a wide embrace. Indeed, many people, especially those persons who don't examine their faith too closely, will never transcend this theological stage of dependence. The risk of this spiritual stasis is that people sometimes remain entrenched, mistaking fixed beliefs for faith, thinking other views of God are wrong, if not evil. The benefit, of course, is obvious—in a changing, often unpredictable world, a theology that asserts God's parental, unfailing care can be a strong comfort to those who suffer, so long as they don't examine their beliefs about God too closely or critically.

Ideally, however, our relationship with God, like our relationship with our parents, changes and matures as we age. Shortly after graduating from high school, I decided against going to college immediately and began working full-time at an energy company. As my income and desire for freedom increased, I moved into my own apartment. While I stayed in frequent touch with my parents, speaking with them several times a week, I no longer required their daily care. This isn't to say I didn't cherish them. Indeed, we remained close and visited back and forth frequently, something we continue to do. But the nature of our relationship had changed from one of dependence to mutual friendship. In the past, my father assigned me tasks and responsibilities, which I felt obligated to do without argument. Now our relationship had shifted, his authority over me had lessened, and my own autonomy had increased.

In some ways, this change was liberating and welcomed. At other times, when I realized my well-being depended on me and not my parents, I felt some anxiety. During difficult times, I thought how nice it would be to move home and let my parents provide for me as they had when I was younger. However, I suspected if I returned home, I would have to surrender a good bit of freedom, which I was unwilling to do. I had grown fond of my liberty and appreciative of the more mutual, collegial relationship I enjoyed with my parents. I also believed it unwise to become dependent at the very time in my life I needed to become independent. While my parents had never seemed weary of caring for me, I suspect if I had returned home, that day would have come.

Around the same time, my theology also began to shift. Previously, I had viewed God as an all-powerful parental figure who regularly intervened to fix my problems. I had no hesitation asking God to involve herself in the minutiae of my life, providing one thing or another to make my life more pleasant. But I began to wonder whether God might be weary of my daily pleadings and might be expecting me to assume a larger share of duties, not only my own, but the duties of those unable to care for themselves. Though God hadn't said, "I will no longer do everything for you. It's time you learned to help yourself and others," I sensed the key to my spiritual growth lay in my willingness and ability to become God's partner instead of God's child. Rather than expecting God to do everything, I would assume greater responsibility for my care and the care of others. I became concerned about the poor and weak, believing it my duty as a human to assist in their growth and progress. I felt especially tender toward those who suffered and believed that God didn't so much want me to pray for them but wanted me to comfort and assist them to the extent I could. In short, I believed the answer to the prayers of suffering people wasn't in a God who intervened, but was found in people who loved and cared.

Now my parents are aging, and our roles have reversed. They rely more heavily upon me and my siblings. Of course, for their sake I wish they were still physically able to do everything they had once done, but that is not likely to happen. The day will come when they will depend upon me and my siblings in daily matters, just as my wife's mother, Ruby, needed our regular assistance.

Similarly, God is no longer able to do for me, at least in my mind, what I once thought him capable of. Consequently, instead of viewing God as one who helps me accomplish my purposes, it is now my joy to help God accomplish the divine purpose—seeking the best for others and seeking the growth of the beloved, which is to say everyone.

When I was young, God was my parent figure. As I moved into independence, I viewed God as my partner, and experienced a mutuality of friendship and responsibility I had never before felt. I wonder if the next step is a return to total dependence on God as I age and weaken, or further independence as my growth in wisdom hopefully accompanies my growth in age. Perhaps then I can be even more useful to God. I hope it is the latter, but I don't yet know. I do know my thoughts about God and suffering have changed and will likely change again as I age. Though I am occasionally anxious about some aspects of aging—the decline of my physical and mental powers—I look forward to my evolving understanding of God, how it might affect me, and whether it will bless others. Perhaps this might also be the answer to the age-old dilemma of suffering and evil. As we mature, we find ourselves looking less and less to a parental God who intervenes on our behalf and instead take more seriously our own roles and responsibilities in a world transformed.

Salvation

When I was growing up, a prominent church in our town split. Like most religious divisions, the reasons publicly given were theological in nature. One faction declared the other faction to be insufficiently Christian; the other party countered with charges of its own, sides were chosen, lines were drawn, and secession followed. One group retained the original building, while the breakaway group met in a rented space before purchasing land on the outskirts of town.

While the reasons given for the schism were theological, those close to the fray said the division had more to do with a long-festering struggle for power. Religious divisions are rarely attributed to petty motives, so the separation was attributed to the noble cause of spiritual purity. The departing group believed the other group lacked sufficient zeal for saving souls.

As is true of most divorces, the departing party worked hard to justify its departure, while those left behind labored diligently to prove their detractors wrong. As a result, our town had two groups of people bent on saving the rest of us—one group stressing the importance of personal salvation, the other not wishing to be accused of spiritual apathy. The upshot was that those of us who belonged to neither church became targeted for conversion.

I don't want to assume religious terms such as *salvation* are universally agreed upon or even known, so before I go further, let me say that those involved in the divided church understood salvation to mean a secure eternity of joy and bliss with Jesus after one's death. This was accomplished by believing Jesus's death on the cross had paid the penalty not only for one's personal sin, but for the original sin of Adam, thereby satisfying God's demand for holiness and justice. God's forgiveness, and Jesus's triumph over sin, were demonstrated by Jesus's resurrection from the grave. One only had to acknowledge his or her personal sin, recognize the separation from God one's sin had caused, believe Jesus had acted on his or her behalf, and accept the gift of Jesus's vicarious sacrifice in order to enjoy that same gift of new and eternal life.

I belonged to a Catholic church at the time, where, while the goal of salvation was the same—to enjoy eternal bliss with Jesus—the means of salvation were assuredly different. There I was taught that my salvation rested upon my participation in the sacraments of the Roman Catholic Church. Because our understanding of salvation differed from that of our fellow Christians, we Catholics were often approached and invited to make Jesus our personal savior. This first happened to me when I was in the fourth grade and a schoolmate who attended the new church approached me at recess, informed me I was going to hell, and invited me to his church the following Sunday so I could listen to their preacher and get saved. By then they had erected a handsome new structure at the edge of town.

I responded with a territorial passion, declining my peer's offer of salvation, sharing my opinion that he was the one destined for hell, then walked away, leaving him to think about his precarious situation. He was not the first to press his narrow understanding of salvation upon me, nor would he be the last. Indeed, these overtures seem to happen with startling regularity, since millions of Christians think it their sacred duty to save others from the ravages of hell, especially those who do not share their beliefs.

The Appeal and Limitations of Narrow Salvation

Today, this understanding of salvation is most often associated with the more conservative, evangelical strains of Christianity. One hears it not only from many pulpits, but also if one happens upon a television preacher, where it is a constant staple of their diet, second only to pleading for money. This worldview persists, and might even be growing, for at least three reasons.

First, it is an easy matter to persuade people that humanity is inherently flawed and in need of rescue. While we usually have a high regard for our personal morality, we tend to believe others, especially those who are different from us, are ethically and spiritually inferior. We are far too quick to suspect the worst of our fellow beings. Ministers who preach about the fallen state of humanity receive a steady chorus of "Amens." Consequently, any theology of salvation that assumes human sin and brokenness is an easy sell. Why we see weeds instead of grass is a mystery to me, but the tendency is real and does extensive spiritual damage to us and others as we look not for the best in one another, but the worst.

Second, this understanding of salvation flourishes because many Christians are theologically uneducated. They are unaware of church history, ignorant of the implications of their beliefs, and ill-informed about the Bible, how it came to be written, and the proper method for its interpretation. Nor have they taken the time or made the effort to learn how to think theologically. This is understandable. People, after all, have families, careers, and other responsibilities that make the sustained study of religion and theology difficult, if not impossible. I was only able to do it because my wife agreed to provide our living for the eight years I was in college and graduate school. I am only able to continue it because I have found a way to wed my interest in theology with the practical need to earn a living for my family. Not many people are able to do that.

I am fortunate to be able to read, study, reflect, and speak about the matters I love. Even then, there is much I don't understand. There are others who haven't had my advantages of time and study who are as passionate about their theology as I am about mine. Though passionate, they have not been able to give their beliefs sustained and careful thought. Sometimes those same people, mistaking passion for knowledge, are the most dogmatic about what the rest of us should believe. I say this with no sense of superiority. It is simply a fact. Theology, like any field of study, is best served when persons have the time and resources to consider a matter deeply over a sustained period of time.

Last, a narrow viewpoint of salvation persists because those who embrace it have convinced others it is the historic witness of the church, that Christians have held this understanding of salvation since the time of Jesus. In reality, this theology has its root in the American revivalism of the mid-to-late 1800s, was popularized by well-known preachers such as Dwight Moody, Billy Sunday, Billy Graham, and others. But it is important to bear in mind this is a Christianity the first apostles wouldn't have recognized. It is a distinctly American theology, whose roots can be directly traced to the national religious revival that followed the American Civil War. Indeed, Dwight Moody was a chief architect of this revivalist theology, which asserted, among other things, that the Civil War was God's punishment for America's sin of slavery. From there, it was natural to insist on America's need for repentance and its return to holiness. Echoes of this theology were heard in President Lincoln's second inaugural address when he said, "Fondly do we hope—fervently do we pray—that this mighty scourge of war may speedily pass away. Yet, if God wills that it continue, until all the wealth piled by the bondman's two hundred and fifty years of unrequited toil shall be sunk, and until every drop of blood drawn with the lash, shall be paid by another drawn with the sword, as was said three thousand years ago, so still it must be said 'the judgments of the Lord, are true and righteous altogether.'" Though Moody and Lincoln couldn't know it, they were articulating a theology whose tenets would sink deeply into the American psyche and grip it still today.

While the apostles of Jesus would not recognize modern evangelical theology, it is also true that there are aspects of my theology the apostles wouldn't recognize. The difference is this—I know that, and am not claiming the authority and weight of the historical church, while most evangelicals are. I believe it is perfectly acceptable, and sometimes even desirable, to expand the scope and meaning of Christianity beyond the boundaries of a particular Christian era, while evangelicals would have us believe the credibility of their beliefs rests in their consistency with history. But it is a fabricated history, a history consistent with their theological priorities, based less on fact and more on myth. It is history as they wish it were, not as it was. Nevertheless, evangelicals have done a masterful job of convincing many people that their theology is consistent with the New Testament church, mostly by making that claim fervently and often.

In summary, the appeal of narrow salvation rests in our willingness to believe the worst of humanity, not the best, a lack of sustained theological education, and the ability of those who hold this ideology to persuade the rest of us that their view is historically grounded. When one becomes aware of these limitations, one is liberated to start thinking of salvation differently, which I will soon invite us to consider. Before I do, let me share a few concerns I have articulated in another book, *If Grace Is True,* and touched on above, but which bear repeating because of their importance.

Central to understanding this narrow view of salvation is God's requirement of shed blood for the forgiveness of sin, or blood atonement. Rooted in the Hebrew tradition and early pagan cultures, it asserts that God (or the gods) was somehow unable or unwilling to forgive and accept us unless blood was shed on our behalf—in the Christian's case, the blood of Jesus, who took our sins upon himself, allowing himself to be punished on our behalf, thereby paying the price for our sin and allowing us to be reconciled with God, spared God's judgment and damnation. Additionally, it asserts our eternal well-being is at risk not just because of our personal sin, but because of our fallen nature as a consequence of Adam's original sin.

The problems with this understanding are obvious. How can the same God who urges us to freely forgive be so resistant to forgiveness that he would demand the spilling of blood in order to pardon others? How can God demand a mercy of us that he himself is not willing to extend to others? This is nothing less than divine hypocrisy, inconsistent with the gracious character of God, who freely and generously forgives.

Another problem has to do with the desirability of one person's assuming the sins of someone else. It's a tempting idea, especially for those who wish to escape the consequences of their poor choices, but it makes no sense, nor is it desirable. The very effort of always taking responsibility for another person's ethical failures inhibits the spiritual growth and moral maturation that person desperately needs.

Yet another problem is the idea that Adam's sin has somehow put our lives in jeopardy. It requires great leaps in logic, the obvious one being the literal existence of Adam, which is, quite frankly, implausible. Not even the author of that story likely believed it to be factually true. It is a myth intended to make sense of the twin mysteries of human origins and suffering. As myths go, it is compelling, which is why it has endured. But it is also limited, failing to take into account other complexities. We are left to wonder how two people who never lived, because of a sin that never happened, condemned the rest of us to eternal damnation.

This is not to say myths aren't helpful. They can be quite valuable, prompting a more creative consideration of important matters. Indeed, the beauty of myths is their multivalence, their ability to speak to us on multiple levels,

depending on our own lives and perspectives. Their intention is to expand our understanding and enrich the human
experience, not arm us with scriptural clubs to bludgeon one
another into theological conformity.

Despite our differing interpretations of biblical myths and
our conflicting definitions of salvation and its means, I do
share with my evangelical brothers and sisters the sense that
all is not right in our world, that we humans are imperfect,
that we can be ethically corrupt, sometimes dangerously so.
It is this realization that causes us to reflect on our need to be
healed from that which hinders our growth and happiness. We
both share the conviction that the world, and many of those in
it, need salvation or healing or wholeness, that we are not as
we could be, that sin or immaturity has exercised inordinate
power in our lives. Whatever we call that state of perfection or
spiritual well-being, many of us recognize the great need for it
and desperately yearn for it, for ourselves and others.

To that end, having considered the limitations of a narrow
understanding of salvation, let us think for a moment about
the universal longing to be whole, then look anew at what it
means to be saved or completed and how it might be possible.

Our Universal Longing

When I was ten years old, my father took my little brother
and me canoeing. Though my brother and I had never
canoed, my father had, though this did not alleviate my fears.
When we left home, I hugged my mother an especially long

time, certain I'd never see her again. It was in late spring, it had rained the day before, and the water seemed high and swift-moving. My perception of the depth and current probably had more to do with my fears than with reality. A mud puddle would have seemed fraught with peril that day.

We were carried on a bus fifteen miles upstream to a grove of trees, where we loaded our canoe and struck out, my brother perched in the front, I in the middle, and my father in the rear. We pushed off into the stream, and within a very short distance, no more than a hundred yards, we came upon the first bend in the river. A tree had fallen into the water and debris had piled up behind it, creating a vortex that seemed to suck us in, despite our fervent but failed effort to paddle the canoe away from it. We crashed into the logjam, flipping the canoe, dumping us and our lunch into the water. The current pushed my brother under the limbs, where he became trapped. I could see him pinned under the water but couldn't reach him.

My father dove down, freeing my brother from the logs and limbs. With a child in each hand, my father made for the shore, where he deposited us in the sand and gravel, then went back for the canoe, pulling it to the bank, even managing to salvage our lunch, which he had wisely packed in a waterproof bag.

Though that happened many years ago, I have retained two strong memories of that day. The first was my father's calm, assured voice, after he had tended to us, saying, "I don't think your mother needs to know about this. It would only worry her." He had taken us canoeing despite her suspicions that the day would end in calamity, if not death.

My second recollection is the dread I felt when I first realized we were going to capsize and there was nothing we could do to prevent it. I remember not only my own fear but also my brother's: he began shouting, "Oh no!" over and over even before we struck the downed tree. When we were carried to the shore by our father, the look on my brother's face was one of unmitigated relief and joy at having been rescued.

Over the years, that memory has acquired spiritual overtones. It comes as no surprise to me that, for many people, salvation has come to mean rescue from peril. When I left Catholicism and became a Quaker, the Quaker meeting I first attended would occasionally sing the song "Love Lifted Me." It begins with the line, "I was sinking deep in sin, far from the peaceful shore." The song culminates with an affirmation of the power of God's love to rescue and redeem the stricken. While I can no longer give my assent to some of the theological implications of the song, I do embrace other aspects of it, particularly the sensation of feeling at risk and the power of love to lift or save us.

Our human longing for rescue might well be a universal impulse, whether we understand ourselves threatened by an inherited sinful nature, an angry deity, ignorance, evil, strife, nothingness, indifference, or confusion. Though we might understand our perils and their origins differently, we nod our heads in agreement when we speak of life not being all it could be, that somehow, in some way, we have not reached our potential, that life lacks the significance we desire. We might speak of the need to find ourselves, or feeling out of step, out of touch, or out of balance. Perhaps we feel unloved, or find ourselves unable to love at the depth we wish. Others might speak in more dramatic terms, believing our lives are at physical, mental, and spiritual risk. Whatever the case, we experience our lives as incomplete or unfulfilled, sometimes radically so, so that we might even begin to question whether life can be sustained. This experience of needing to be saved or made whole is not a conservative or liberal one. It is a human longing, part and parcel of our essence. *While this longing for salvation transcends specific religions, it might well be the collective yearning that inspired the invention of religion, if such a complicated phenomenon can be traced to a single factor.*

For many people, indeed, perhaps for most people, the effort to become saved or whole or well is primarily a religious or spiritual effort. They believe that God, however God might be understood, provides the necessary insights or resources to help them on their way. This movement toward salvation or wholeness is often the gateway through which we move toward fuller, more meaningful lives.

Almost every Sunday, our Quaker meeting will have several first-time visitors. More often than not, when I engage them in conversation, they will volunteer why they are visiting, usually in these very words, "I am looking." If time permits, and I sense they don't mind my inquiring, I'll ask what they are seeking. The responses have included community, meaning, understanding, a fresh start, relationship, acceptance, love, or heaven. Behind almost every response is the hope for a fuller, richer life.

What Is Salvation and How Do We Get There?

When I was a child, a mentally disabled man lived several blocks away with his brother's family. Once every week or so, I would visit him, curious about his condition and wanting to talk with him. For reasons I never knew, he was unwelcome in his brother's home, but had been given a corner of the garage along with a narrow bed, chair, and television. He ate alone, was under strict instructions not to leave the garage, and had few visitors, apart from the neighborhood children who would occasionally stop to see him. I now suspect his brother had promised their dying parents he would care for his impaired brother but did so in a manner that required the least amount of work and care.

I've since forgotten many of our interactions, except for one, which has stayed with me. I had just finished delivering my newspaper route and had stopped by to see him. Christmas was nearing, and I asked him what he wanted. He thought for a moment, then said, "A friend. I want a friend."

Because of my immaturity, I didn't appreciate his plight, though I gradually became aware of the relational poverty of the man's life and how, already hampered by a mental disability, he was further plagued by a scarcity of meaningful and nourishing relationships. As I became conscious of these important dimensions to life, I returned to visit him but learned he'd passed away.

I don't know if he got the friend he wished for, but I do suspect if he were still living, and had the ability to reflect, he would equate salvation and wholeness with friendship, it being the one thing he wanted and needed most.

I know a man who is blessed with friendships but struggles with an increasing dependence on alcohol. Though he has sought treatment and devotes much time and energy to remaining sober, he occasionally succumbs to the lure of alcohol and begins anew the cycle of relapse, regret, and recommitment to sobriety. Were we to ask him what salvation means to him, he would probably talk about being freed from the temptation of alcohol and its destructive hold on his life.

Another acquaintance grapples with his inability to sustain a marriage beyond a few years. For him, being saved would mean becoming the kind of person who is capable of an enduring love.

A woman I know worries constantly about money. Though well-off, she is anxious about whether she will have enough for her old age. She has sought counseling, but still feels crippled by her concerns. Salvation to her might entail a broader understanding of what it means to be secure.

Yet another friend works in a job he finds meaningless. Urged by his father to assume control of the family business, he now wishes he'd resisted his father's request and become the teacher he'd always dreamed of becoming. Because he is bright and affable, he has led the company well, but he is unfulfilled and wants desperately to take up more satisfying work. For him, salvation would mean pursuing a vocation that utilizes his gifts, brings him joy, and blesses others.

This is all to say that there might be as many definitions of salvation as there are human needs, that the specific requirements for wholeness vary for each of us, depending on which things keep us from being well and whole and healed. For the rich young ruler who approached Jesus, asking the path to eternal life, it meant giving his wealth to the poor. For others of us, it might be something else entirely.

Meteorologists have coined the term *perfect storm* to describe what happens when weather fronts intersect to create an especially powerful storm. Similarly, there are times when the facets of our lives join together and we are fully alive. We find life meaningful and rich. We are in right relationship with others. We love deeply, and we are loved. We have a keen awareness of the Divine Presence and feel ourselves profoundly connected. We experience feelings of bliss. It can happen during worship, while in nature or service, or in intimacy with our spouse or partner. They are moments of intense and profound connectedness, when we might say, "This is what it means to be fully alive! This is how life should be!"

Gratitude and joy permeate our lives. We experience a clarity of purpose. Our work, our vocation, which before might have felt burdensome, now feels worthwhile, our yoke light.

The mentally impaired man who lived down the street from me, despite his intellectual shortcomings, had a knack for repairing lawn-mower engines. They would be pushed into his garage, belching smoke and misfiring, then would roll out a few days later, their engines harmonized and humming, perfectly tuned. Perhaps this is analogous to what we in the church have typically called salvation. *Though I once believed salvation involved my affirmation of doctrines about Jesus, I now believe salvation is that condition when our heart, soul, strength, and mind are so attuned to the Divine Presence and to one another that we are empowered to love, relate, think, and act. But it is to love, relate, think, and act toward a specific goal—the growth and betterment not only of the self, but of others, indeed even the growth and betterment of the enemy.*

Jesus, rather than being the *means* of salvation through his sacrificial death, was instead the *archetype* of salvation, embodying this unity of purpose and divine receptivity, and in that power he lived as one transformed, even as he was transforming the lives of others.

To be honest, I have not yet discovered the secret, if there is one, to experiencing this receptivity and empowerment each moment of my life. My life, like most lives, is fragmented. I enjoy moments of deep clarity and belonging when I feel united to God, others, and creation. But I also undergo periods of spiritual dislocation, when I feel estranged from God, my fellow humans, and the wider world. Though I have heard others describe salvation as a constant state of spiritual bliss, and perhaps for them it is, I have found salvation to be a land in which I sometimes dwell, while at other times it has seemed a faint and distant memory.

When people visit our Quaker meeting and say they are seeking, I suspect what they are seeking is this depth of belonging, this bliss. In those moments, I try never to oversell salvation by promising them a constancy of elation I have not known myself. Rather, I invite them to a life of divine awareness so that when God's Spirit, which truly does blow where it will, comes their way, they are well positioned to acknowledge and enjoy it and in those moments know salvation.

The Paradox of Salvation

Though I have dedicated much of my adult life to the work of the church, there are some aspects of its life with which I am uncomfortable, the chief one being its certitude when a little doubt was in order. Nowhere is this certainty more present than when the church has spoken of salvation. We not only know who is saved—if not specifically, then at least theoretically— but we have refined the formula by which salvation happens and can replicate the exact scenario, as if God were a well-trained deity awaiting our command to fetch and heel.

My experience is otherwise. While I believe God delights in the divine/human intersection, it seldom happens predictably. I attend meeting for worship each Sunday, hoping to experience a moment of connection and many times have. But I have also left our meetinghouse feeling as if the hammer missed the anvil, that no spiritual sparks were generated. I don't believe God was playing hard to get or making herself absent so our hearts would grow fonder. Rather, I believe, like all moments of transcendent beauty, our experiences of salvation are unpredictable, not dependent on a set of theological laws, but more on timing, whimsy, sensitivity, and the intersection of random events. Let me offer an example.

Near my home is a meadow of wildflowers that I walk through several times a week the year round. Sometimes the meadow seems drab: the sky is overcast, the grasses and flowers bent and brown, the trees bare, and the air cold, damp, and uncomfortable. Other times the meadow is radiant: the flowers are in full bloom, the sky is blue, the temperature is ideal, and the sun is shining, striking the blossoms at an angle that makes them appear even more colorful and vibrant. A variety of factors—the sun, the plants, the time of day, the wind, the climate, the season—conspire to make that meadow a paradise.

While I cannot control those variables, I can place myself in a position so that I am more likely to experience their intersection. This is why I regularly and frequently visit the meadow, to increase my opportunities to witness such loveliness. When I speak of salvation as a paradox, this is what I mean. While we cannot demand that God be known to us at a specific time to bless us in a particular way, our opportunities to experience salvation, our chances of being wholly united with God, others, and creation rise as we "visit the meadow" or cultivate our sensitivity to God's presence.

What is the consequence of these encounters? What happens when we experience the Divine Presence? We are more adequately equipped to commit our hearts, souls, minds, and strength toward the divine purposes of growth and betterment, the growth and betterment not only of the self, but of others, indeed even the growth and betterment of our enemy.

This is why, even though I no longer believe some of the more traditional assertions of the church, I continue to participate in its life, just as I continue to visit the meadow. I want to be a witness to beauty. I want to expand my opportunities to encounter the Divine Presence. I want to feel more deeply connected with my fellow beings and be more a part of the surrounding creation.

I place myself among others, I walk the meadows, because I no longer want to be just an "I" but an "us." Perhaps this is what the monk had in mind when he was approached by a well-intentioned Christian who asked him if Jesus was his personal Savior.

"No," the monk answered, "not my personal one. I've always preferred sharing him."

What if this were salvation, so that even as our hearts, souls, minds, and strength are growing, so are others'; all of us moving together toward holy beauty and perfection, transforming ourselves and delighting God? This is an evolved Christianity I could happily and wholeheartedly embrace.

Prayer

When I was confirmed in the Catholic Church as a boy, I was given a Bible with a zippered pocket on the back cover that contained a rosary. My grandmother had ridden the bus from southern Indiana to witness this event, and afterward, when we were at home and Sunday dinner had been served and eaten, my grandmother sat on the couch and summoned me to her side.

"Do you know what this is?" she asked, holding the rosary.

I replied that it had something to do with prayer, but I wasn't sure what. (My interest in religion had not yet blossomed.)

"Let me show you," my grandmother said, taking my hand in hers so we could finger the beads together. She showed me how each bead represented a specific prayer, then repeated each one aloud with me, her fingers moving from bead to bead, keeping track of our recitations.

Like I did, my grandmother learned the rosary at an early age, and until her death at the age of eighty-eight had prayed the rosary several times each day. One of the enduring memories of my childhood is my grandmother, seated on a bench in their living room, near a small statue of the Virgin Mary, her rosary in hand, her gaze distant, her lips moving in prayer. She began each day praying the rosary, concluded lunch with more prayer, then prayed the rosary again before she went to sleep. Indeed, as she lay dying, her fingers moved instinctively along the beads of an imaginary rosary.

My family took my grandmother's prayers quite seriously. We believed that because of her piety, God heard my grandmother's prayers and was faithful to grant them, so we would regularly write or phone her to ask her prayers for certain endeavors. When I received my first job at the age of sixteen, I was certain my grandmother's prayers were responsible, and for several years thereafter, whenever I faced a significant challenge, I would phone my grandmother to ask for her intercession.

Though I can't know the extent to which my grandmother's personality was formed by her habit of prayer, I do know she was a sweet, kind woman. While her prayers lacked originality—I had heard her pray countless times but not once could recall her offering a spontaneous prayer—they seemed to possess a quality of power. Given my youth and the totemic nature of the rosary, I came to believe prayer had a magical element to it, that its efficacy depended not only upon the faith of the one praying but upon his or her strict adherence to the method of prayer.

Even after I left the Catholic Church to become a Quaker, prayer seemed to me to be a magical undertaking. For years I believed prayer could persuade God to do something she would not otherwise have done, except that numerous persons had prayed. While I was serving as a youth pastor, a teen in my Quaker meeting was diagnosed with cancer. I set about recruiting as many people as I could to pray for him, believing there was a tipping point, that if an unknown but significant number of persons prayed for my young friend, God would hear our prayers and intervene to save his life. I remember thinking if I were to persuade one more person to pray, my friend would be cured.

During this period, I read books about prayer, asked more mature Christians to share their experiences of prayer, and even came to believe effective prayer hinged on a secret insight I had yet to discover, that if revealed could produce miraculous results. At one point, I became convinced if I prayed at my friend's hospital bedside, holding his hand while I prayed, he would be healed. I drove 130 miles one way to pray at my friend's bedside and was happy to do so, believing it would effect his cure.

Despite my efforts and prayers, and the labors and entreaties of many others, the young man died, and with it my certainty about prayer and God.

I write this now, some twenty-plus years later, not to elicit your sympathy or defend my theological journey, but because I suspect each person reading this book has experienced similar sorrow and spiritual disquiet. Each has prayed faithfully and diligently that a friend, child, spouse, or parent would be healed, only to stand by the loved one's still form and weep.

To say that our prayers for the young man's healing did no good would be untrue. While they didn't accomplish our hoped-for end, they might have provided us with a sense of comfort and connection. The opportunity to vocalize our deepest fears in prayer is helpful, and perhaps even psychologically beneficial. To be the recipient of loving prayer can itself be healing and restorative. Our young friend with cancer seemed deeply moved by our prayers and said on numerous occasions that they gave him strength and peace.

Some people, though not receiving the desired result of their prayers, can still appreciate their benefits. Others are so wounded by what they perceive to be the pointlessness of prayer, they dismiss any of its potential advantages and conclude that God is uncaring or impotent, or doesn't exist. Still other persons, fearful of acknowledging the apparent inability of God to effect change, resort to religious platitudes rather than face the doubts raised by unanswered prayer. I recall when parents I knew lost their child to a virulent form of cancer, despite much fervent prayer by their church, family, and friends. During the child's funeral, the minister remarked, "God always answers our prayers. Sometimes he says no." That observation, while intending to diminish the sense of loss and betrayal the parents were experiencing, had just the opposite effect on me. For it implied that if God had the power to say no, God also had the power to say yes. Studying the parents' reaction, I sensed they had drawn the same conclusion. Shortly afterward, they left the church, wanting nothing to do with a God who, in the church's cosmology, had the power to heal their daughter but had decided against it.

Though I didn't realize it at the time, the varied reactions to unanswered prayer were unfolding before my eyes:

Some persons, though not getting what they had asked for, were appreciative of unanticipated blessings, such as patience or compassion or the sense of community that often accompanies collective prayer.

Some persons simply accepted the outcome as God's will and didn't question it, believing God's will was a mystery not to be understood by mere mortals.

Some, having been taught God answers the prayer of the faithful, lost their faith in God or grew angry with God or feared God or outright rejected God.

As a pastor, I have witnessed each response, have sympathy for each reaction, and have concluded that most people don't lose their faith because they experience suffering, but because the church's explanations of their suffering is ultimately intolerable.

It is time to state the obvious—at any given moment, millions of desperate and hurting people are asking God to exercise his power toward a particular end and God isn't doing it, either because God can't or God won't. Our efforts to explain God's inaction only deepen their resentment and compound their confusion. In light of this, something must give—either our understanding of God and prayer must change, or we must acknowledge prayer as an antiquated practice and abandon it, lest it cause more hurt and pain than it resolves.

It was the crisis of prayer, more than any other factor, that motivated the change in my theology. The images of God I had been taught were no longer sustainable. The idea that God, moved by the intensity or quantity of prayer, would reach down from heaven to alter earthly circumstances, fell apart when I witnessed the volume of unanswered prayer. To be sure, some prayers were "answered," but never in numbers sufficient to prove prayer's efficacy. I began to suspect other things I'd been taught about God were also inaccurate. Of course, once I realized that, everything came undone.

When Religious Systems Collapse

The disintegration of one's belief system can be frightening, for some people even terrifying, which is why many religious people fear inquiry. But I was not scared—not due to any personal courage, but because I wasn't old enough for my theology to have taken root, and having so little invested in it, I found the prospect of moving to new spiritual ground exciting. But for those who are heavily invested in a given theology and faith, rethinking matters they once thought settled can be perplexing.

Whenever I speak on spiritual matters, it is not unusual for someone to raise his or her hand and ask, "Has prayer changed for you?" It is usually an older person, typically someone for whom the spiritual dimension of life is very important. The person usually goes on to say that since his understanding of God changed, he no longer knows how to pray. He often seems nostalgic for what prayer once meant to him. Sometimes he confesses to feeling guilty for not praying as often as he once did. He might have memories of his parents praying regularly, just as I remember my grandmother and her prayer life. He's wistful, remembering the simple devotion of others, and says, in one form or another, "I wish I could pray like I used to, but I no longer can."

I ask him whether, if given the chance, he would return to his earlier beliefs, but he doesn't see that as an option, saying it was too difficult to sustain a belief in the God of his earlier years. He does, however, seem to miss the uncomplicated prayer that undergirded that period of his life.

"How do you pray?" he asks. "Tell us about your prayer life."

One of my earliest memories of prayer, apart from learning the rosary from my grandmother, happened when I was ten years old and wanted a bicycle. For years I had ridden bicycles passed down to me from my older siblings or cobbled together from bicycle parts I'd found at the town dump. Our family was large, my parents' income modest, and new bicycles weren't a priority. But in my tenth year, around August, I began actively campaigning for a bicycle for Christmas. In addition to reminding my parents of my wish, I also believed my chances for a new bicycle might be helped if I asked God for it. So every night, for several months, lying in my darkened bedroom, I prayed to God, requesting a bicycle.

I was very specific, not trusting either my parents or God to get the details right. I wanted a green Schwinn Racer with drop handlebars. Unbeknownst to me, my parents had contacted my grandfather, asking him to purchase and store that very bike at his house until the week of Christmas, it being rather difficult to hide a bicycle from a curious boy. As Christmas drew near, I daily inspected the basement, attic, and barn for a bike and even searched my neighbor's home, growing more frantic each day, assailing the gates of heaven with my request.

A few days before Christmas, I ventured to the basement in search of presents, and in the farthest corner of the cellar, in the coal bin behind the furnace, I noticed a blanket draped over an object I hadn't seen before. Pulling aside the blanket, I discovered a gleaming, green Schwinn Racer.

To say I believed, in that moment, in the efficacy of prayer would be an understatement. I was hooked on prayer, believing I had discovered the secret to earthly wealth and happiness. I fell to my knees before the bike, thanking God for his generosity. It didn't occur to me to thank my parents. Having been told so many times they couldn't afford it, I naturally assumed God was responsible.

This became the pattern of my prayer through my adolescence and early adult years. *I believed the purpose of prayer was to acquire a blessing I could not otherwise attain except through God's intervention.* Thus my prayer life became especially active during end-of-semester testing or when I wanted a date with a particular girl, needed money to buy a car, or desired some other advantage I believed beyond my reach through ordinary means.

While my prayers were not always successful, I avoided enough catastrophes to make me think they were at least partially helpful. I prayed for As, and got Bs. The cheerleader wouldn't date me, but another sweet girl consented to spend time with me. I couldn't afford the Camaro I wanted but did manage to buy a Volkswagen Beetle. While my requests were not met exactly as I had prayed, they were close enough; I was content and so continued to believe in the value of prayer. Indeed, my early years were so generally positive that it was easy to believe God was bending an ear toward my prayers and altering circumstances to bless me. With this evidence in hand, I prayed regularly and often, freely directing God to do first one thing and then another, all of which seemed to happen or at least occur to a degree sufficient to sustain my faith in prayer.

This understanding of prayer persisted until my mid-twenties, when the teen in my youth group was stricken with cancer and died. I described that event above, noting that while the prayers of many people weren't answered, they might still have found that experience of prayer beneficial. But for me it was a desolate period. For the first time in my life, I was unable to redeem or make sense of a tragedy. When I spoke with others of my spiritual disillusionment, they encouraged me to consider what I had learned from the experience—compassion, the importance of life, the peace that comes with prayer. I replied that I could have learned those lessons without my friend's dying. One woman said the experience taught me God was not my spiritual Santa Claus, hastening to satisfy my every request, to which I responded that taking the life of a young man to teach me that lesson when God could just as easily have denied me a bicycle was especially senseless and cruel.

Though this event did not lessen my interest in God or church, it did change my prayer life. Quite simply, I stopped praying. More accurately, I stopped praying as I had. I no longer prayed in the evening before going to bed or in the morning after rising. Indeed, I usually only prayed in times of great anxiety, when I thought my wife or children were at risk. Even then, it was done spontaneously, without conscious thought, and I would catch myself doubting, even as I prayed, whether my prayers actually made a difference.

I also continued to pray publicly as a minister but was careful not to ask God to do something for me or anyone else that couldn't also be accomplished through skill or work.

Otherwise, I would be forced to wonder why God answered some prayers and not others, and that seemed to me an imponderable question. I never asked God to heal anyone, but would instead ask God to show us how to love and help those who hurt and suffered. If pressed to ask for God's intervention in a specific manner, I would invite the person to join me in "Quaker prayer," take the person's hand, and pray silently, so as not to betray my doubts about the effectiveness of intercessory prayer.

My Gradual Return to Prayer

During this long stretch of doubt, I met many pastors who were described by their congregants as "prayer warriors" and envied them their certainty and faith. One pastor told me to "just believe," and I desperately tried, but couldn't. Then one day, perhaps ten years after my young friend's death, I came across a quote by Meister Eckhart, the thirteenth-century German mystic, who said, "If the only prayer you ever say is thank you, that would be sufficient."

For reasons I still don't understand, that saying galvanized me. I could say, "Thank you." Gratitude required no suspension or compromise of logic or reason. I determined to find one thing in each circumstance for which I could be grateful. So when people were sick, I was grateful for doctors and medicine and hospitals. When people were lonely, I was grateful for friends and pets and community. When people fought and killed and murdered, I was grateful for kindness and forgiveness and those who worked for peace. When I was hungry, I was grateful for farmers and migrant workers and chefs. But curiously, I seldom directly thanked God for anything, except in my public prayers when I felt I was speaking less for myself and more for the community of faith to which I belonged.

Then, early one morning, while walking in the woods near our home, two birds burst out of a bush ahead of me. They were brilliantly colored Baltimore orioles. I had never seen that particular species before but had heard they were becoming common in our region of the country. The two birds flew to a nearby tree and perched there, side by side, like miniature suns.

Without hesitating, I said aloud, "Thank you, Lord, for such beauty."

I studied the birds until they flew away, then resumed my walking, thinking about my spontaneous prayer. Why had I thanked God? Though I had cultivated the habit of thanking people for their contributions to my life, I had not been as effusive with my praise when the source of joy and beauty was less certain. Should I thank natural selection for making the birds so beautiful and be grateful to Charles Darwin for bringing that process to our attention? Should I thank the publishers of *The National Audubon Society Field Guide to North American Birds* for helping me identify the birds as Baltimore orioles? No, for while those people had made me aware of the birds and their origins, they had not created them.

Though I had grave doubts about God's willingness or ability to intervene in our lives at our behest, I had never stopped believing in God's penchant for beauty. Nor did I find it implausible to believe God might have kick-started the whole creative enterprise (though I can't help but wonder who kick-started God). Divine creativity made as much sense as any other theory I had ever heard. But more than that, as I aged and encountered all types of people, I noticed the happiest people I met had one trait in common—they were all profoundly and extravagantly grateful. Even when they weren't sure who to thank, they knew it was important to have a spirit of appreciation. I wanted to be like them, and since there was no else around to thank that morning in the woods, I thanked God.

So after nearly twenty years of stilted, infrequent conversation, God and I began to converse again. Not like before, when I would make requests and imagine God hastening to grant my desires. But more spontaneously, usually prompted by some beauty or joy, when I would thank God or whatever source I thought responsible for the gift I was enjoying.

After a year or so, I found myself lying in bed, at the close of each day, retracing my steps, recalling the day, thanking God for all the wonderful things that had happened to me or others. I did this silently, so as not to disturb my wife. I would fall to sleep with thoughts of gratitude. Had someone asked me at the time, "Do you pray?," I would likely have said no, though now I wouldn't hesitate to attach that word to what I was doing.

Interestingly, during all my years of silence, I never fretted about my lack of prayer. I assumed if there were a God and that deity cared about integrity, she would not be grieved by my silence but would understand it and perhaps even appreciate my reluctance to pray insincerely. During that time I heard many people equate one's spiritual status with prayer, but I never felt any less a Christian, reasoning that spirituality was a journey and that one day I would find a way to communicate with God consistent with my experience of her.

I likened it to when I was a teenager and for several years found it difficult to speak with my parents. I certainly didn't doubt their existence. I only found it difficult to have any substantive discussion with them that didn't leave me feeling even more estranged. The fault was not theirs, for they were, and are, gracious people. The fault, as I reflect on it, wasn't even mine. My yearning for freedom simply outstripped my capacity for self-reliance. I resented my dependence, even as I sensed its necessity, and every conversation reminded me of my continued need for them just when I longed for liberation. Eventually, as I matured, our relationship warmed without any sense of rancor or recrimination on anyone's part.

So it was with God. When our conversation resumed, I felt no need to apologize for my silence or seek forgiveness. I perceived God understood the reason for my quiet, was

pleased to reconnect—if, in fact, we had ever been discon-
nected—and had much love for me, as she did for all people.

Prayer was no longer a laundry list of requests but was
instead a heightened awareness of God's presence in creation,
in others, and in myself. I resolved to simply pay attention to
my life, to better discern the Divine Presence and to respond
gratefully to that Presence—and that became my prayer.

Prayer as Paying Attention

When our children were toddlers, it was our custom to take
them on a walk around our neighborhood each evening.
Before long, we were a familiar sight, and people would often
stop us to see our children. Of course, not all our neighbors
were outgoing, and though some offered a timid wave, they
seldom spoke to us or approached us. We never dismissed
them as snobbish, rude, or cold, realizing they were intro-
verted. We would smile, call out a greeting, and keep moving.

But one woman, whenever she saw us, would walk to the
end of her driveway to intercept us. She quickly learned our
names, the names of our sons, their birthdates, and gifted
them with small toys on their special days. This continued
for several years, and during that time we grew close to her.
Though our sons were too young to realize it, the greatest
gift she offered them was her attention. But my wife and I
knew it, and our appreciation for her awareness and consideration grew.

Because our children are important to us, we want others to value them too. When people go out of their way to notice and greet them, we are especially grateful and believe we have found a special friend. When I served with an older pastor, he advised me to bend down at eye level to each child with whom I spoke. "Parents notice how their children are received. Don't ever ignore a child."

Of course, we all notice whether we're being paid attention to, whether others notice us or dismiss us. There are few experiences as alienating as being overlooked or rebuffed. One of the most compelling traits of Jesus was his attentiveness, his keen sensitivity toward the lonely, ostracized, and hurting. This theme of attentiveness runs through the life of other God-bearers too. We believe their relationship to God is especially rich and comprehensive because of their awareness of others. They notice what others overlook.

When I first began serving as a pastor, I had the opportunity to be present at a talk given by a well-known speaker. Coincidentally, I happened to be near the door when she arrived and was being greeted by those in charge of the event. One of the hosts took the speaker by the elbow and guided her to a private room so she wouldn't be swamped by well-wishers. As they passed me, I overheard the host say to the speaker, "There's a man here who is a real annoyance. I'm going to steer you away from him so he won't dominate your time."

The speaker asked where the man was, and the host discreetly pointed him out.

I followed the gaze of the speaker and noticed a man sitting by himself, avoided by everyone.

The speaker slipped from the grasp of her host, walked over to the man in question, sat beside him, and began engaging him.

I was aware of the man's story. He had lost his wife several years before and had become something of a pariah since then because of his emotional neediness and his tendency to glom onto others.

The woman placed her hand on his shoulder, and they began visiting. I couldn't hear what was being said, but he was speaking intently and she was paying rapt attention to him, to the annoyance of the hosts, who wanted to extricate her from the situation. Several minutes passed before she hugged the man, offered him a word or two of encouragement, then shook his hand good-bye. By then, I was close enough to hear her say, "It was an honor to meet you."

In my half century of living, it is no exaggeration to say I have heard, read, or given tens of thousands of prayers. Some private, others public. Some of them simple, others elaborate and flowery. Some heartfelt, others insincere and meant for show. But I can't honestly recall any prayer as eloquent and timely as the one I witnessed that day. For what is prayer but our attentiveness to the Divine Presence, and what was that speaker's holy observance but her recognition of the overlooked pain of another?

Though I didn't appreciate it then—it took several decades

for me to understand the scope of her grace—I now realize that while prayer comes dressed in many forms, attentiveness underlies each one. *Indeed, the root of prayer is attentiveness to the Creator and the created.* For true prayer does not consist of words so much as it consists of one heart turned and tuned to another.

Praying Without Ceasing

Though raised in the church and familiar with Bible stories from an early age, I never seriously engaged the Bible until my late teens. I approached Scripture, more accurately the New Testament, reverently, intending to follow it as closely as I could. One morning, while reading the fifth chapter of 1 Thessalonians, I encountered Paul's counsel to the church to pray without ceasing. Given my understanding of prayer at the time, I wondered how that was possible. Prayer was done on one's knees, or in a similarly deferential posture, and required our complete mindfulness. I wondered how one could pray without ceasing while attending to the necessary transactions of daily life, but this did not keep me from trying. I would clear my mind and begin praying, reverting to one-sentence prayers when distracted, which was frequent, given the restlessness and short attention span of teenage males. But I soon grew discouraged and came to believe praying without ceasing was a gift enjoyed by spiritual giants like Paul, unachievable by the rest of us.

But as my understanding of prayer shifted away from ritualized supplication toward a state of gratitude and attentiveness, I began to reconsider what it meant to pray without ceasing and wondered if unceasing prayer, when understood in that context, was now a real possibility. Is it possible to orient one's self to the world in a state of gratitude and attentiveness? Of course. Indeed, not only is it possible, it is the secret to a changed self and ultimately a changed world.

What if this became our prayer? What if every person we encountered received the fullest measure of our attention, with no regard for her status or what she might offer us in return? What if we learned to listen deeply and attentively to one another, undistracted by the many diversions that pull us away from one another? What if, in every moment and circumstance, we were filled with gratitude, able to appreciate the blessing or lesson each situation presented? What if this became our habit, so that we embodied Paul's encouragement to pray without ceasing?

Now let us suppose our heightened attentiveness to others made us more acutely aware of the Divine Presence, not only in others, but in ourselves—in fact, made us so deeply conscious of the divine will that we became living testaments to grace and transformation. Would we not then understand that prayer is so much more than ritualistic posturing or pleading requests, that it is instead a loving and living dialogue, steeped in gratitude and consecrated by attentiveness, able to help us fully appreciate not just God's power, but our own? And understanding our own power, might we then be equipped to answer the prayers of others?

For me, then, this is prayer: that I would in every circumstance be grateful, and in every moment be attentive to others, so that the Divine Presence in me might speak to the Divine Presence in others, and in the speaking, grace might flow and flower.

10
The Church

Like many people, I retain memories of my first day of school. I was very nervous, as were most of my fellow students. To ease our fears, our teacher, Mrs. Wilma Mann, introduced herself, asked each of us our names, then encouraged us to share with the rest of the class some information about our families. When it was my turn, I told the class I had a mother and a father, four siblings, and a dog named Zipper. The exercise had its intended effect. As each of us stood to speak, we visibly relaxed and before long felt right at home in Mrs. Mann's first-grade class.

For the first twenty-three years of my life, whenever I was asked to describe my family, my answer was the same. I had a mother and a father, four siblings, and a dog named Zipper. (My family, though creative in a number of ways, is unimaginative when it comes time to name a dog. For the past four generations, all our dogs have been named Zipper.) A few months after I turned twenty-three, I married my wife, Joan.

Then, when asked to describe my family, I said, "I'm married to Joan." Depending on the circumstances, I might also have mentioned my father, mother, and siblings.

Eight years after we were married, my wife and I had our first son, then three years later our second son was born. Today when someone asks me to describe my family, I respond by saying, "I'm married to Joan, have two sons, Spencer and Sam, and a dog named Zipper." If I'm in the mood to talk and time permits, I might speak about my extended family, my nieces and nephews, my aunts, my cousins, and perhaps even mention the four Quaker brothers named Gulley who first settled in America in the late 1600s. If I'm feeling especially philosophical, I might even talk about the interconnectedness of the human family and thereby identify as my kin all the peoples of the world.

This is all to say that over the span of years, my description of family has changed, as the circumstances of my life shifted. The same has been true with my understanding of the word *church*. For the first fifteen years of my life, church meant Mary Queen of Peace Catholic church in Danville, Indiana. It meant Father Edward McLaughlin celebrating the Mass each morning and a stream of nuns instructing me in the catechism of the Roman Catholic Church.

Though that was a limited understanding of church, it was sufficient at the time, especially since I wasn't interested in religious matters and viewed church as inconvenient if not outright objectionable. I knew about Protestants, since my father's family were Baptists and I had Methodist and Quaker friends, but I was no more interested in their churches than I was in my own.

At the age of fifteen, I refused to attend church any longer. My mother, realizing forced worship was counterproductive, said, "You are old enough to decide what faith you will have, or whether you will even have one."

By "faith," I understood her to mean which Christian denomination I would belong to, since there were a notable lack of Buddhists, Hindus, Muslims, and Jews in our small town. For a year I didn't attend church, then was invited to participate in a Quaker youth group, and within a few years joined the Society of Friends. During that period, my understanding of church began to expand. As a consequence of my movement into another denomination, I became aware of the wider Christian family and began discussing with my friends and acquaintances what it meant when one was "in the church."

Eventually, I would include in my understanding the nearly 39,000 Christian denominations now in existence, and might even speak in broader terms about "communities of faith" so as not to overlook the other religions that populate our world. For I see in all those groups the same desire to understand, belong, love, grow, and honor the Divine Presence.

I imagine the same is true for you, that over the years your understanding of the church has changed dramatically. Like me, your earliest perception of the church was probably linked to a specific congregation but widened as your life and worldview expanded. This is not to say your appreciation for your local church diminished, only that your awareness of and appreciation for the larger spiritual world increased. Your world grew, and as it did, so did your understanding of spiritual community.

Just as my perception of the church's breadth has changed, so, too, has my understanding of its mission. I once believed the purpose of the church was to urge people to accept Jesus as their Savior so they could go to heaven when they died. As I've noted elsewhere in this book and others, that belief requires theological assumptions I can no longer affirm.* Consequently, my understanding of the church and its mission has changed. This did not happen overnight, nor are my current perceptions necessarily final. If I continue to be part of the church, which is a strong likelihood, my views on the church will no doubt change, given the evolving nature of life.

Curiously, as my perception of what it means to be Christian has expanded, the influence and role of the church has diminished, at least in the United States. While more people seem drawn toward spirituality, they no longer believe the church is its sole purveyor. Indeed, they often see organized Christianity as the enemy of spirituality, as evidenced by the

---

* I have dealt at some length with the issue of atonement theology in *If Grace Is True,* a book I cowrote with James Mulholland, published in 2003 with HarperOne.

oft-heard statement, "I'm spiritual, but not religious."

How can the church, in a world of diverse faiths, best honor the priorities of Jesus while helping to create a world made new?*

A Possible Way Forward for the Church

In his book *The Future of Faith,* Harvard educator Harvey Cox describes three distinct eras of church history. The first era was the initial three hundred years of the Christian movement. Since the biblical canon and creeds were not yet formed and codified, the early church focused on following the teachings of Jesus. Cox refers to that period as the Age of Faith. It was marked by a countercultural commitment to the ethos of Jesus, even to the point of suffering persecution. Though politically powerless, the early Christians were noted for their passion for the priorities of Jesus, his regard for the outsider, and his commitment to peace. The ripple effect of his life and witness were still very much present in the fledgling Christian community. A number of Jesus biographies, four of which were eventually canonized, circulated throughout the early churches, generating enthusiasm for this rabbi's teachings.

The next era Cox refers to as the Age of Belief, which

---

* For a more detailed reading on my view of the church, please see *If the Church Were Christian: Rediscovering the Values of Jesus,* published in 2010 by HarperOne.

stretched from the fourth through the mid-twentieth centuries. This second era was less about emulating Jesus and more about believing the right things about Jesus, focusing on orthodoxy and correct doctrine. During this period, Christianity gained political power, but often at the expense of its witness to the radical nature of Jesus's life and message.

The third era, which Cox calls the Age of the Spirit, has been unfolding for the past fifty years. It is a movement away from dogma and formal religion, toward the integration of diverse religious expressions and spirituality.

Most of us are old enough to have witnessed the gradual transition from the second age to the third. While this ecclesial evolution has been at times painful and difficult, it has also been liberating. The reins of religion, once held by a centralized hierarchy, are being taken up by a wide and unaffiliated array of spiritual leaders, including women. But this era is not just about diffused power. It is also about the church's loss of power. The diminishment of the church's stature occurred for several reasons—the lack of a unified Christianity and a universally agreed upon leader, the loss of the church's moral stature because of spiritual, ethical, and sexual abuse by some of its leaders, and the multiplication of spiritual movements that better fulfill the spiritual needs of those who seek a transcendent dimension to life.

It is helpful to take up each of these causes individually:

Institutions become powerful for one of two reasons. They either have at their helm a charismatic individual who is able to generate enthusiasm, enact change, and solidify broad and durable support; or the institution, and its attendant traditions, are so well established that people assume it is the only viable means by which something can happen.

In regard to the church, the Protestant Reformation made it possible for us to imagine churches taking other forms. We discovered, after nearly fourteen hundred years of a Roman Catholic monopoly in the Western world, that it was entirely possible to be fully Christian outside the Roman Catholic tradition. The world didn't end, we didn't go to hell (at least we hope that's the case), and we still experienced the Divine Presence.

We also realized others outside the apostolic chain of authority could provide effective leadership. While this was a generally positive insight, the presence of so many leaders within the church necessarily limited their authority and reach. The United Methodists don't feel inclined to heed the pope, nor do Roman Catholics feel compelled to follow the president of the Latter-day Saints (Mormons). There is not a single charismatic church leader, nor is there likely to be, who can galvanize the entire church into action or speak for the church with any sense of broad authority. Even if such a figure were to arise, another Martin Luther type, he or she would meet stiff resistance from the established church, for charismatic people invariably promise change, which is the last thing entrenched institutions want, preferring placeholders and gatekeepers instead. These factors make dramatic, broadly accepted leadership nearly impossible, crippling the church's ability to speak with one voice, thereby diminishing the church's influence.

Add to that dilution of power the scandals across the church involving the outrageous behavior of prominent televangelists and the horrific incidents of abuse within the Roman Catholic Church, which, though they represent the acts of a minority of priests, have been handled so poorly that many people now scoff at any notion of the church's moral authority. *Because moral standards and spiritual influence are so closely linked in the minds of most people, no religion that violates the former will enjoy the latter.* Hence, the world's largest Christian denomination has experienced a tremendous loss of credibility, which has spilled over into the wider Christian community.

Alongside its diminished influence and loss of moral weight, the church must compete in an age when a variety of spiritual movements provide a transcendent dimension to life. Christianity is no longer the only game in town. This is best illustrated by the headstone emblems approved by the U.S. Department of Veteran Affairs for deceased soldiers. For many years the markers were limited to some version of the Christian cross or the Star of David. Today, some thirty-nine symbols mark the headstones of fallen soldiers, with more being added each year. Those emblems now include, among others, the Muslim's crescent, the atheist's atomic symbol, the Wicca's pentacle, the Hindu's om, and the Baha'i's nine-pointed star.

The church's monopoly in America and the West is over. This isn't to say Christianity will die, only that it is now one player among many attempting to win the hearts and minds of the next generation. The Age of the Spirit that Harvey Cox identifies, marked by a diverse and lively array of religions, is fully upon us. No longer, when someone says she is a person of faith, can we assume the object of her faith is Jesus.

It is tempting to blame the erosion of creedal and institutional Christianity on those beyond our walls, to fault outside influences for the tremors that have cracked our institutional foundations, just as fundamentalist Muslims have blamed the West for the perceived decline of traditional values in their own communities. But life is never that simple, and the reasons for change seldom that clear. As much as Christianity has been altered and influenced by movements beyond us, we have also been shaped by circumstances within the church, largely by progressive Christianity and its embrace of modernity, tolerance, acceptance, and openness. Some Christians have thus concluded that we are our own worst enemies, that our best option for a viable future lies in our determination to embrace a rigid faith in order to stave off the adulterating influences of other cultures and religions. But I would contend that this has been tried repeatedly throughout our long history and always ends the same—in suspicion, intolerance, exclusion, division, and, finally, war.

No, if the church has a future—indeed if our world has a future—it will rest in the church's ability to honor and assimilate the best of each religious tradition, just as Jesus found virtue in Samaritans, publicans, centurions, and Gentiles.

How this good man came to be the focus of a creedalism that ultimately excludes others is a mystery for the ages. *The incorporation of other traditions into our own will undoubtedly change us, but for the better, for it will lead us toward one another, which is also and always a movement toward the Divine Presence and the universal grace that Presence represents.*

To be sure, if one believes Christianity is primarily about worshipping Jesus, a faith that incorporates other religious traditions will be considered heretical. But if one believes Christianity is primarily about following the example of Jesus, then it is easy to imagine a faith informed by men and women of goodwill, though of diverse traditions. If the future of the Christian faith is creedalism and believing the right things about Jesus, then other traditions will be viewed as the enemy at worst, or contaminants at best. It will be a return to the Age of Belief, and in that sense a spiritual regression. *But if the future of the Christian faith is about taking the best from each tradition, while helping people negotiate their spiritual journeys with grace and dignity, then the church might well inspire a world made new.*

Let us think for a moment of the benefits of assimilation. My maternal grandfather's family, the Quinetts, moved from Belgium to the United States in the early 1900s. After settling here, my grandfather met my grandmother, also of Belgian origin, and they were married. My grandfather kept largely to his Belgian community, a group of glasscutters living in Vincennes, Indiana. But his daughters made their way into the larger world, went to college, married, and had children who did the same. Between my grandfather and his brother, their children and grandchildren include a doctor, a pharmacist, several nurses, a university employee finishing his PhD, a musician, a writer and pastor, a business-school student, an elementary-school teacher, a respiratory therapist, a teacher of developmentally disabled children, a school principal, a prosperous investor, three general contractors, a business manager, and a real estate manager.

My family's story is not unusual. It has been repeated by millions of American families over the years, helping our nation flourish. Countries and cultures who welcome and assimilate people, ideas, and insights grow and prosper; countries and cultures that don't, stagnate.

This is no less true of churches. *The future of Christianity will rest in our ability to make our spiritual boundaries more porous, welcome the wisdom of other faiths, and borrow the best from other spiritual traditions, even as we share with them the stories and insights of Christianity.* This in no way dishonors the contributions of Jesus, but recalls his appreciation for those persons thought to be outside the circle of God's favor. When searching for an example of faith, he lifted up a Roman centurion. When illustrating compassion, Jesus spoke of a despised Samaritan who stopped to help. His willingness to see the good beyond his own tradition is a clear reminder for us to do the same.

Christianity, from its very start, was an invitation to believe God was at work in the wider world, far beyond the parameters of any one religion. When the church has forgotten the expansiveness of God, it has descended into a narrowness of mind and a meanness of spirit. When the church has remembered, it has been a light to the world and a balm and blessing to hurting people everywhere. *To celebrate the life and witness of God in other faiths is not to diminish Christianity, but to elevate a core conviction of its namesake.* This, I believe, should be the work and witness of the twenty-first-century church. This expansiveness of spirit will not be reached by our rigid adherence to orthodoxy, but by our willingness to enter into spiritual community with others who, though they might believe differently, still embody a high regard for the Divine Presence who enlivens and embraces us all.

The Church as Model Community

If we understand the Bible as its writers hoped, as a collection of stories, letters, parables, poems, proverbs, and wisdom sayings intended to convey their emerging understanding of the Divine Presence, then we should pay close attention not only to its overt points, but also to its minor themes and undercurrents. One such theme in the Gospels is the precedence Jesus assigned to community. His earliest priority was to invite people to share life with him, seeking out individuals with whom he could enjoy relationships. Though the church has often said he was "making disciples," I suspect his motives were less theological and more practical. Like most of us, he valued friendship and community, sensing his life and theirs would be enhanced and expanded by close relationships.

It is not uncommon to hear people speak dismissively of churches as "social clubs." Sadly, it is often a charge hurled by one Christian against another. It is usually meant negatively, intending to call into question a church's lack of zeal and orthodoxy. For years, I said such things myself, criticizing the social dimension of other churches, which I believed had a detrimental effect on the church's true mission of saving souls. I no longer believe that. Indeed, if the activities of Jesus were an indication of his priorities, it is clear he placed a high value on the social aspect of spirituality and faith.

Commonly known as the Golden Rule, Jesus's counsel to treat others in the manner we wished to be treated was understood to be the summation of Israel's Law and the prophetic tradition (see Matt. 7:12). Think of that—the distilled wisdom of rabbinical and early Christian teaching concerned itself with our treatment of others. What is that, if not a standard for community? *Much of Jesus's ministry concerned itself with the proper treatment of others, linking our gracious treatment of others to our spiritual well-being.* The argument could be made that for Jesus, religious community was to be first and foremost a prototype of appropriate social community.

Just off the square in the small town where I live is a simple brick Unitarian-Universalist meetinghouse whose outdoor message board is regularly posted with thoughtful sayings. One of my favorites was coined by a well-known Unitarian theologian and teacher, James Luther Adams, who said, "Church is a place where you get to practice what it means to be human."

The first time I read that sign, I had just received a scathing letter from a fellow Quaker who, upset by my theology, had written to urge me to leave the church. Years before, the lady writing me and I had enjoyed a warm friendship, but as my theology became a focus of concern, she grew upset with me and our friendship cooled. I was saddened by our frayed relationship and also angry and disappointed with my friend, believing she had grown spiritually rigid.

On a morning walk, I experienced a jumble of emotions as I passed the Unitarian-Universalist church and read Adams's words on the message board—"Church is a place where you get to practice what it means to be human."

Adams's insight rang true. Now, finding myself in a contentious and painful situation, I had been presented with the perfect opportunity to practice what it meant to be human. When I reached the town square, I stopped to sit on a bench. It was early in the morning and no one was about, so I sat undisturbed and reflected on the situation. Specifically, I tried to see the matter from her perspective. I had called into question her long-held spiritual values and had contributed to dissension in the community of faith we both loved. Though I believed many involved had acted rashly, I was also not without fault. I had at times spoken hastily, giving little consideration to how my remarks might be received.

I resolved that while I had an obligation to speak honestly about my emerging beliefs, given my role as a leader in the church, I also had a responsibility to do that tenderly and thoughtfully; in that regard I hadn't always succeeded. In that moment, I understood the reason for her anger and vowed to treat her with kindness and warmth in hopes of restoring our frayed ties. I had no sooner decided that than my anger toward her dissipated, replaced with a deep appreciation for our long years of friendship. Not long afterward, I saw her and greeted her warmly, and she responded in kind.

I am not certain my change of heart would have occurred outside the context of church. For as I sat on the town-square bench, my mind was filled with stories and examples of forgiveness I had learned at the church's knee. While I belong to a number of organizations and institutions, only the church has given me the language of reconciliation and has concerned itself with my human growth and betterment. When I have been angry, it has taught me to forgive. When I have been lonely, the church has provided friendship. When I was happy, it celebrated with me. When I was sad, it shared my grief. When I was egotistical, thinking only of myself, the church corrected me and taught me to consider others. When I was stingy, it taught me generosity. And when I was fearful, it taught me courage. In short, the church let me practice what it meant to be human. Not just any kind of human, but the best human I could be.

It is also abundantly clear that what the church has provided for this Christian has also been provided to the Jew, the

Muslim, the Buddhist, and others by their spiritual communities. All of them, in their own contexts, have been taught what it means to be human. To be sure, I and others have not consistently lived up to the ideals of our spiritual communities, but those ideals are no less important and imperative.

The Church as Initiator of Reconciliation

It seems obvious that if our world is to endure, spiritual communities will have to seriously address religion's culpability in a fractured world. It is also clear that unless religions take the initiative in reconciliation, universal peace and goodwill are unlikely to happen. Christianity, with over two billion worldwide adherents, is well placed to model the kind of inclusive, transforming spiritual community I and many others have envisioned. Given the priorities of our namesake, it is possible to reframe the "Christian story," so that even as we remain faithful to the witness of Jesus, we can simultaneously affirm the value in other religious expressions. It is in no way unchristian to honor and replicate Jesus's willingness to see good in those traditions outside his own. Why can't we appreciate the Muslims' discipline, when they abstain from food for long daytime hours during Ramadan to better focus on their relationship with Allah? Why can't we be grateful that Orthodox Jews honor God with a Sabbath day, and why can't we adopt that same pattern of rest and renewal? Why can't we see the wisdom of Native American spirituality that honors the earth and teaches unity with creation? Why must we continue to act and live as if Christianity is the sole source of spiritual truth and wisdom?

For Christianity to lead the way in a movement of reconciliation is doubly appropriate, given how many of the world's religious battles were instigated by governments and movements within the Christian tradition. As a collective act of historical repentance, it is fitting that Christianity initiate the spiritual reconciliation of the world's religions. *If we, in an effort to prove our ecclesial purity, claim to be rooted in the early church, if we claim a direct spiritual connection to those Christians who've gone before us, then aren't we also obligated to make amends for our predecessors' excesses, especially when those excesses have harmed others?* Having enjoyed the tangible benefits of our ties to the historic church, aren't we also compelled to make amends for its sins?

Not only would such conciliation be Christlike, it would save us from Christianity's more unfortunate alliances. For too many years, at least in the United States, Christianity has been too closely allied with civic and temporal power. One seldom hears the word "God" without the words "and country" following closely behind. This has often been detrimental to both our religious and national interests, when the values of one have compromised the principles of the other. A Christianity whose focus is unity among the world's religions and peoples could more readily divest itself from a civic relationship that hasn't proven all that beneficial and has more often than not corrupted the church's integrity and mission. The very effort to unify the world's religious movements and direct its considerable energies toward a good end might in the end have a purifying effect on Christianity.

This would necessarily be accompanied by a shift in the church's concerns. The focus of Christian worship, especially since the fourth century, has been sacramental, doctrinal, and conversional, urging people to believe certain doctrines about Jesus, affirming those beliefs through the repetition of ceremonies and creeds, while also attempting to convert those persons outside the Christian fold. In terms of increased growth and power, this approach has been successful. But in terms of fostering a faith the world can live with, the verdict is checkered. When our goal has been the recruitment of others to our way of thinking, we have often lost our way, valuing recruitment above reconciliation.

Our passion for recruitment lies in our desire to have others make the same religious choices we have made, thereby confirming our wisdom and good sense. In that sense, recruitment is a self-centered activity, valuing others primarily for their willingness and ability to confirm our decisions. But a church centered on reconciliation, not re-cruitment, begins with the assumption that others are our equal partners in loving work, not targets for our evangelism. When that is the case, we will no longer view those outside the church as mistaken, confused, spiritually lost, or damned.

Instead, we will see in them the very potential and promise Jesus saw in those he encountered.

The next Christianity will offer an understanding of the Divine Presence the whole world can live with. This will not happen in a generation. Just as the Age of Belief unfolded over sixteen centuries, this path of reconciliation and assimilation will take time to accomplish. *But our first step is long overdue—our simple admission that the church holds no monopoly on truth, that the God who made all the world has graciously created other spirit people and Jesus-types to embody this message of peace, compassion, and unity.* Again, this in no way diminishes the stature of Jesus but acknowledges his importance by recognizing God's desire to replicate his gifts and virtues in others around the world and throughout history.

While I am not yet sure how this next Christianity will unfold, I know it must if the world is to endure. For in a world rent by division, our collective impulse toward the divine might well be the synthesizing factor necessary for the world's continued evolution. Just as the world can no longer afford cultural and political discord, neither can it long endure the escalating tension of religious division. A significant number of thinkers believe that for the world to have a future, religion must be dispensed with altogether. They believe, and their evidence is compelling, that religion has done more to harm world unity than foster it. If our only option is religion as we have known it, I would agree. *But a third way exists—the decision to affirm and practice the best of each spiritual tradition, caring less about which religion birthed the principle and more about its contribution to a wise and gracious humanity.* That tradition would care less about doctrinal purity and more about a world restored, where all people are God's people and all things are made new. Whether the wide family of Christianity will be a part of that movement remains to be seen, though it does seem clear that the same Jesus who welcomed the outsider would hail a gracious and unified world as great, good news.

11

The Afterlife

Before I went to college, I worked several years at an electric utility with a man who was a Jehovah's Witness. He was a kind man and I was theologically curious, so we had many conversations about spiritual matters. The Witnesses, I learned, have some rather curious beliefs, among them that only 144,000 people will go to heaven. When my friend told me there were several million Witnesses, I surmised that my chances for making it to heaven via the Witnesses were slim, so I lost interest. Because the goal of Christianity seemed to be the attainment of heaven, I saw no need to embrace any faith that couldn't guarantee my future happiness.

Eventually, the concept of heaven lessened in importance to me, and today I give it little thought. That might be a minority position since many Christians continue to believe the purpose of Jesus and the Christian faith is to prepare us for an afterlife of eternal bliss, or conversely, if we aren't Christian, an afterlife of never-ending misery. While that construct no longer makes sense to me, I do believe our ideas about the afterlife are vitally important, given their ability to influence our treatment of others in this life. *Indeed, I would go so far as to say that the moment we embrace a theology that consigns some people to hell, we have given ourselves tacit permission to treat them like hell.* Therefore, I prefer to talk about the reasons for, and ramifications of, heaven and hell, which can be known, as opposed to talking about their actual existence, which can't be known.

Why Heaven and Hell?

Anthropologists and sociologists have long considered the role and origins of religions. We know they've arisen for a variety of reasons, mostly as attempts to explain profound mysteries, such as suffering, evil, death, and the seeming injustices of life. But not all the mysteries we experience are negative. We also feel the need to explain the profound joy we experience, our mystical encounters with a transcendent Presence, the beauty of creation, the wonder of birth, the order of the natural world, and our feelings of love and belonging.

Of course, as we have learned more about our natural world, much of what was once mysterious and inexplicable is now understood. For instance, we now know the cause of eclipses and no longer interpret them as portents of evil or divine wrath. Though still amazing, we now understand the concepts of reproduction and birth. To a large extent, we are now able to figure out weather patterns and are even able to predict future weather with some reliability. While a large number of humans are suffering at any given moment, we are familiar with the reasons for their distress and have even been able to alleviate their suffering, though right now we lack the ability to eliminate their afflictions altogether.

Even though we haven't solved all of life's mysteries, it is clear we have made huge strides in knowledge, enabling us to better understand our world, making us less likely to appeal to God to manipulate events or protect us from that which we did not understand. Indeed, our advancements in understanding our natural world have been so significant that it seems to me the new frontier for human exploration lies not in the realm of physical science, but in the sphere of human emotion, motive, and behavior. Why do we act the way we do, especially when our behavior is so obviously personally destructive? How can human behavior be moderated to eliminate war, evil, greed, and injustice? What causes two siblings, raised in the same environment, to respond differently to similar nurturing and stimuli? Why does one become a beloved teacher and the other a drug addict?

When the earliest humans gathered around a campfire, perhaps they asked themselves similar questions. Perhaps their earliest efforts in social control was the realization that reward and punishment could be useful tools in reinforcing good behavior and discouraging poor behavior. Good behavior was rewarded with esteem, respect, social status, trust, and ultimately eternal blessing. Poor behavior was punished with rejection, shunning, imprisonment, censure, and ultimately eternal condemnation.

As social controls, heaven and hell were powerful tools. The promise of heaven made our earthly burdens bearable, promising those who suffered on earth a future of heavenly blessings. They only had to endure their present misery with docile acceptance, good cheer, and faith in order to enjoy God's riches in the next life. Indeed, were I a member of the ruling class, I could do nothing better to guarantee my continued power and prosperity than to convince those under me that God had placed them in their position, that God willed their situation, but that one day they would be blessed with the riches of heaven for their faithful acceptance of God's will, namely, subservience to me. It is telling that hymns of heaven and future glory have invariably risen from the impoverished and enslaved classes, the rich and powerful having no need to sing of future blessing.

As for hell, were I poor and powerless, unable to challenge injustice in this life, I would embrace a theology, and preach it endlessly, that those who abused their power and neglected the poor would one day face God's judgment and wrath for their tyranny and indifference. I would write a story about Lazarus and Dives, in which the poor man, Lazarus, enjoyed the cooling refreshment of heaven, while the rich man, Dives, languished in hell. I would do so, hoping this would curb the excesses and evils of the rich and powerful, that it would cause them to notice my suffering and act with haste to help me and those I loved.

We must face the possibility that our earthly lives might well be our one chance for awareness, consciousness, and existence. It is therefore incumbent upon us to create heaven on earth for those who have known only hell. This was the heart of the prayer Jesus taught his followers, to work for God's kingdom to come on earth. Our labor and longings were to be invested on this side of life, not on an afterlife we have no proof exists. Indeed, I am increasingly convinced that Christianity's historic focus on an afterlife has done more harm than good, distracting us from the real goal of God's reign.

Heaven and Hell: The Great Distractions

Early in my life as a minister, I was invited to pastor a small, rural congregation not far from my home. Because I was new to ministry, my superintendent offered a few words of counsel, including the advice not to make any dramatic changes until I had been there for at least a year. "Let them get to know you first," he said. "That way when you suggest changes, they'll trust you."

I thought it was a wise suggestion, so devoted my first year to getting to know my congregants. Within a short time, I noticed they had a curious fascination—one could almost say an obsession—with the afterlife. Their hymns, Sunday-school lessons, meal conversations, and Scripture readings focused inordinately on going to heaven. On the Sunday of our church's homecoming, when they welcomed back former pastors and members, they spent nearly two hours singing songs about heaven, working themselves into a state of near ecstasy. They lived in constant fear that one of their number would do or think something so evil it would jeopardize their own chances to enter heaven, so misbehavior or perceived heresy was met with swift punishment, including banishment from the church.

So deep was their preoccupation with the afterlife that any encouragement I made for them to focus on helping the hurting in this life fell on deaf ears. Simple requests to gather food for the hungry were met with indifference and even scorn. I recall one man, responding to my suggestion we help feed the hungry, saying, "Let us give them spiritual food." His remark was met by a chorus of "Amens." The same man, after discovering an elderly couple in the congregation were living together without being married, was adamant that their "sin" could very well cause God to send the entire congregation to hell.

As I got to know the people better, I discerned their fixation with the afterlife stemmed from the teaching of a pastor who had preceded me by several years. That pastor preached each Sunday, for nearly twenty years, on the one thing about which he felt reasonably certain—heaven. His texts were old hymns, which described in exaggerated detail the particulars of heaven—streets of gold and gates of pearl, angels, harps, and heavenly beings. For the reasons described above, this appealed to many in the congregation, and the entire time I was there I was often encouraged to preach about heaven.

Curiously, the congregation had scant interest in spiritual growth, self-actualization, significant biblical study, addressing human need, or welcoming others into our church community. Once, out of deep frustration and a desire to connect with them, I preached a sermon about heaven, describing in rich and vivid language its theoretical glories. The message was roundly applauded, and I was congratulated by many for finally being "a real preacher."

"Maybe the Lord's gonna use you after all," one man told me.

So immersed was the congregation in the afterlife that every effort I made to excite their passions for bettering this life was met with resistance. I would talk of peace and they would say, "There'll never be peace until Jesus returns and takes us to heaven." I would speak about the poor and hungry and the necessity of helping them only to be told "the poor will always be with us." In short, their preoccupation with tomorrow's heaven served as a distraction, crippling their desire and capacity for Christian practice today.

The congregation was small, around fifty people on any given Sunday, so the fate of the world wasn't hanging in the balance. *But when one recalls the vast stadiums of people who've gathered to watch prominent evangelists reap a harvest for heaven, when one considers the billions of dollars and hours spent to prepare people for a hypothetical afterlife, one can't help but wonder whether those resources couldn't have been directed toward more tangible needs in this life.* While at that church, I began to realize heaven and hell were profound distractions to authentic Christian living, diverting our attention away from vitally important work. In this sense, the church's traditional emphasis on the afterlife isn't benign. It has real and dramatic consequences, causing us to lose sight of the very life Jesus called us to live. For that, and other reasons, it behooves us to examine more closely the church's misguided focus on heaven and hell.

Besides deflecting our attention away from more pressing matters, the church's focus on the afterlife, and the certainty with which we speak of it, presents another dilemma. *Speaking with such confidence about something that can't be proven damages our credibility.* It is one thing for us to share our experiences of life's transcendent dimension. Almost every culture, race, and generation speaks of some mystical encounter with the Divine Presence, though the names they attach to that Presence differ. But to extrapolate from those ethereal experiences an entirely different plane of existence that occurs after we die is intellectually suspect. For the truth is this: as much as we might wish for an afterlife of bliss and reconciliation—I wish it myself—we have no tangible proof such an afterlife exists. While I, and many others, have been intrigued by anecdotal evidence of near-death experiences, they do not occur with sufficient consistency and regularity to be considered authoritative. *Consequently, to construct an entire religion around the attainment of a theoretical afterlife calls into intellectual question every other claim made by that religion, including those claims for which some proof does exist, such as the life of Jesus and his values, which ought to be the primary focus of Christian life.*

It's Time for Honesty and Grace

If there are hallmarks of the Christian life, they are honesty and grace. When we confuse hope with fact and create theologies that condemn those unlike us to hell, we fall short of both ideals. If we must have an afterlife, if we simply can't imagine the Christian faith without it, then let it be one in which all people everywhere are gathered. But let us be modest about it. Let us speak of it not with certainty, but with hope. And if our lives are already blessed, let us be slow to claim it for ourselves, lest we become greedy for favor. Instead, let us apply ourselves to creating heaven on earth for those who suffer, and if we fall short, then let us hope God in her goodness has created another level of bliss of which we are unaware, so that in the end others will enjoy the goodness we have known. For if there is a heaven, but it is only for us, that would surely be hell.

An Altar Call

12

From time to time, I visit churches whose theology is more evangelical than my own. Occasionally, near the conclusion of worship, the pastor will give an altar call, an opportunity for those who've been moved by the sermon to come forward and dedicate their lives to Jesus. The goal is always the same—to ensure the respondent has acknowledged his or her estrangement from God and "accepts Christ as Lord and Savior." What it means to accept Jesus as your Lord and Savior has not always been carefully explained by those pastors, though I realize, and perhaps they do, too, that an understanding of discipleship evolves over one's life.

While my own Quaker tradition has not made altar calls a priority, I have come to appreciate the powerful role of public affirmation and find myself wishing those of us in the progressive wing of Christianity had a similar rite. For I believe the life and virtues we are calling people toward—peace, justice, mercy, reconciliation, enlightenment, and transformation—are worthy of public declaration and celebration. While our world hasn't always been well served by religion, it has been enriched by our human inclination toward spirituality and our sincere efforts to bring wholeness and healing to broken persons and situations. The public affirmation of that longing is important, if only to remind ourselves and one another of the higher life to which we are called.

One central aspect of Christian fundamentalism is the intense promulgation of their message these past hundred years. Their message has been widely and passionately spread, permeating our culture, giving them the appearance of orthodoxy. While the worldview of progressive Christianity is deeply rooted in Judeo-Christian history, we have not kept pace with our conservative brethren in the effort to articulate our message. We don't spend billions of dollars a year on radio and television programs to persuade people to join us. We don't believe those who disagree with us are lost or condemned. We don't send missionaries forth, encouraging people to abandon their spiritual heritage and embrace ours. We don't believe the purpose of the state is to legislate or enforce our religious beliefs, nor do we believe societies are best served by rigidity of thought, whether it is religious or political. Though our convictions are rooted in Scripture, they have often been expressed by a minority voice, overshadowed by more authoritarian images of the Divine Presence. It has consequently been an easy matter for Christian fundamentalists to counter our images and insights with biblical examples of their own that reinforce divine judgment and wrath. It is no wonder so many people struggle to discern what it means to be Christian.

Though I have never seen a study verifying this, if anecdotal and experiential evidence can offer us any insight, it seems to me a great number of progressive Christians didn't start that way. We grew up in religious traditions much more conventional than the ones in which we eventually settled. In my years of pastoring and writing, I have met many people who've said something along these lines: "I grew up in a fundamentalist home . . ." They then describe their spiritual journeys to their present faith, which is invariably more progressive than their childhood faith. Again, there are exceptions to this observation, though I have noticed the trend often enough to suspect it is common.

Our earliest experiences are often deeply imprinted in our subconscious minds. This is no less true of our religious and spiritual experiences. Even though we chose to move beyond them, traces of them still linger, still exert an influence, and still inform our decisions and perceptions. I remember, in my early twenties, while in a headlong rush toward Christian fundamentalism, having an encounter with the Divine Presence that was at once exhilarating and frightening. It was exhilarating because it opened my life and mind to a different understanding of God, a God who loved far beyond the parameters of Christianity. But I was also frightened, because I couldn't help but wonder, despite the beauty and grace of my encounter with the Divine, that I might be mistaken. In the faith of my childhood, the consequence of wrong thinking was dramatic—eternal damnation. When I eventually concluded that God was honored by my noblest thoughts, not my fears, my worries dissipated.

Perhaps as you've read this book, you've also experienced the mixed feelings of exhilaration and fear. The traditional views of God, Jesus, and self no longer resonate with you, nor are they consistent with your spiritual experiences. But even as you evolve toward what you perceive to be a better Christianity, you have doubts and anxiety, wondering if your current course pleases God. Rest assured, God is always pleased by our determination to think high and noble thoughts. It is, of this I am convinced, impossible to think too generously of God. How often God has born our mean and narrow thoughts of her, how often God has struggled to move us beyond our miserly ideologies, which only belittle her character or reputation. How pleased God must be when our minds consider the broad expanse of God's compassion, creative energy, and commitment to our evolution.

God is creating a better Christianity. God is doing it right now, through people just like you, who refuse to worship a cultural deity whose compassion extends no further than the horizon. While this better Christianity encompasses the priorities of Jesus, he has not been the only prophet to articulate its life-giving virtues. Indeed, I have found it wherever people live with dignity, grace, and compassion. Because this better

Christianity is not mindful of boundaries, it has no special need to be linked to any one denomination or religion. It will happily bear the name of Islam, Judaism, Hinduism, or one of any thousand possible names, for it gladly recognizes in others the Divine Presence common to us all. It labors not for its own proliferation, but for our spiritual and human evolution, and it will not rest until all are loved and all can love. It is hope eternal; it is grace unending. And God is creating it through you, right now.

Questions for
Small-Group Discussion

Chapter One: The Evolution of Faith

1. Think for a moment of your earliest church experience, then reflect on your present relationship with the church. Has your understanding of what it means to be Christian changed over the span of your life?

2. Why do you suppose religion and science have not cooperated with each other? Do you believe they will always be mutually antagonistic, or can you envision a cooperative effort that furthers the aims of both institutions? If so, what might that look like?

3. It was noted that significant shifts in the church's structure are almost always accompanied by or inspired by a theological change. What do you think might be the next significant

shift in the church, and how will our theology change to defend and support such a change?

Chapter Two: Revelation: On Knowing God

1. Do you believe God has ever spoken or revealed something to you? How did that feel? Did it affect your life?

2. Are you skeptical if people tell you God has spoken to them? Why or why not?

3. How do you discern whether a revelation or insight is from God?

Chapter Three: God

1. Can you recall your earliest image of God? How did that early understanding affect your spiritual life in later years?

2. Is it possible to make final conclusions about God and still remain open to spiritual enlightenment?

3. If asked, could you provide a one- or two-sentence definition of God?

Chapter Four: Jesus and Jesus-Types

1. Do you believe it diminishes or emphasizes the importance of Jesus to believe God might have created other God-bearers or Jesus-types?

2. Is the value of Jesus to be found in his uniqueness or in his openness to the Divine Presence, or both?

3. Is the divinity, or nondivinity, of Jesus important to your understanding of Christianity? Why or why not?

Chapter Five: The Living Spirit

1. Have you ever experienced what you believed to be the Holy Spirit? If so, how would you describe that experience.

2. In your experience, what are the indications of the Holy Spirit's presence?

3. Why do you suppose our leadings or insights from the Holy Spirit differ so widely? Why does the Holy Spirit seemingly cause one group of Christians to do or believe a certain thing while simultaneously leading another group of Christians to do or believe just the opposite?

Chapter Six: Who Are We?

1. Can you recall the first thing the church taught you about yourself? What effect did the church's understanding of humanity have on your life? Does it affect you still?

2. Has it been your experience that we tend to live up to or down to the expectations others have of us?

3. Is your current self-perception generally negative or positive? What role did religion have in forming your self-perception?

Chapter Seven: Suffering and Evil

1. What has been your experience with intercessory prayer? When you have prayed in faith for someone to be healed or

for an evil to be averted, were your prayers answered to your satisfaction?

2. Have you ever known anyone whose disappointment with God caused him or her to stop attending church? What were you able to say to that person? Have you ever been disappointed with God?

3. How have your perceptions of God affected your understanding of and response to evil and suffering?

Chapter Eight: Salvation

1. What is your definition of salvation?

2. Do you believe humans are intrinsically depraved and in need of salvation? If so, how long have you believed that, and who taught you? Is it consistent with your experience of the people you know?

3. Do you believe salvation is possible because of Jesus, or do you believe Jesus exemplifies what it means to be saved? Perhaps you have another understanding of Jesus and his role in salvation. If so, can you explain it?

Chapter Nine: Prayer

1. How would you define prayer? Have you found prayer helpful? Have you ever been disappointed by prayer?

2. Is prayer easy for you? Difficult? Has your understanding of prayer changed over the course of your life?

3. I believe the root of prayer is attentiveness to the Creator and the created. Is prayer as attentiveness a helpful understanding for you?

Chapter Ten: The Church

1. Do you believe the following is true: to celebrate the life and witness of God in other faiths is not to diminish Christianity but to elevate a core conviction of its namesake?

2. Will the church, as it presently exists, be a force for good or ill in the future? What benefits do you believe the church has to offer the world? What are the shortcomings of the church?

3. Do you believe Jesus intended to start the church?

Chapter Eleven: The Afterlife

1. If there were no afterlife, would you still participate in a community of faith?

2. Do you believe the concept of an afterlife has been a help or hindrance to our spiritual well-being?

3. Has the possibility of an afterlife ever affected your treatment of others? If so, how?

Chapter Twelve: An Altar Call

1. Have you ever made a public affirmation of your faith? Did it inspire you to a life of deeper commitment?

2. Do you believe God is creating a better Christianity? If so, how might God be doing that through you?

3. Do you believe there is a link between spiritual growth and human evolution?